MW01518884

Cover. Compugraphics by Pierre Bertrand

Edgard Pisani

An Old Man and
the Land

Nine billion people to feed.

Nature and rural societies to protect.

Translated and edited by Paul Perron

LEGAS

New York   Ottawa   Toronto

© 2005 LEGAS No part of this book may be reproduced in any form, by print, photoprint, microfilm, microfiche, or any other means, without written permission from the publisher.

Library and Archives Canada Cataloguing in Publication

Translation of: *Un vieil homme et la terre*.
Includes bibliographical references.
ISBN 1-894508-71-85

1. Agriculture and state--European Union countries.
2. Agriculture--Economic aspects--European Union countries. 3. Sustainable
agriculture--European Union countries. I. Perron, Paul II. Title.

HD156.P5813 2005   338.1'094   C2005-900652-8

For further information and for orders:

http://www.legaspublishing.com

LEGAS
P. O. Box 040328   3 Wood Aster Bay   2908 Dufferin Street
Brooklyn, New York Ottawa, Ontario   Toronto, Ontario
USA 11204   K2R 1B3   M6B 3S8

Printed and bound in Canada

The question of subsistence is too
Vital, too heartrending to be strictly
A social question.

Jean Jaurès[1]

There is only one error and one crime:
To want to enclose the diversity of the
World within doctrines and systems.
Stephan Zweig[2]

Montaigne

---

[1]French politician, philosopher, historian (1859-1914) who was one of the leaders of the Socialist Party. He founded the newspaper *l'Humanité* and as a parliamentarian led the socialist cause. Violently opposed to colonialism and to war, he was assassinated by a nationalist.

[2]Novelist and essayist of numerous works he was deeply affected by the progress and the victories of the Nazis and committed suicide with his second wife in 1942.

Table of Contents

Foreword:

Edgard Pisani

In the *Foreword* to the last book published by Edgard Pisani *A Personal View of the World: Utopia as Method* (2005) I gave an overview of his remarkable career as resistance fighter, civil servant, parliamentarian, Senator, Minister under the two Presidents of the French Republic: Charles de Gaulle and François Mitterrand who most deeply marked France during the last half of the twentieth century. In so doing, I focused more on his contribution to French and world politics, his impact on events that were transformational on the national and international scenes. In brief, I attempted to highlight his role and place in modern history.

Born on October 9, 1918 in Tunis, Tunisia and raised in a traditional catholic family, Edgard Pisani was educated in a classical French secondary school where he chose the option *Philosophy*. He then enrolled in the famous Lycée Louis le Grand in Paris, where he studied to prepare for the entrance exam to the École Normale Supérieure, the most prestigious school in the Humanities in France. Caught up in the resistance movement he was captured, interned and escaped on June 6, 1944; then made his way to Paris where he participated in the Liberation. By chance, he was in the office of the Chief of Police who had been appointed by de Gaulle. The Chief was called to a meeting of the National Resistance Council and ordered Pisani to guard the telephone. He had hardly left when German patrols and armed vehicles began to attack the station. As he was responsible for the telephone he answered and took decisions, sometimes without consulting the three leaders of the police resistance movement and refused to abandon the flag that was flying over the police station. He then accepted a cease-fire that was transmitted by the Consul General of Sweden. After nightfall, all the major commanders of the police station returned and the Police Chief asked Pisani to tell him what had happened during his absence. After having finished, he got up to leave and the Chief asked him to be his Executive Assistant. He met general Leclerc and received the Legion of Honor at the age of 27 for acts of war.

Two noteworthy events occurred during his period of duty as Executive Assistant. One day a guard brought him a calling-card signed Robert Brasillach who asked to see him. After some hesitation Pisani accepted; a tall, lanky, tired man entered, sat down and said that he had had enough of playing hide-and-seek with the police, especially since he claimed he had not betrayed his country. The Executive Assistant was not certain but what he did know is that post-war justice was extremely expeditious. He told his would-be prisoner that he should wait and be judged in calmer times. He informed him that the door behind led downstairs and outside and told Brasillach that he would leave for ten minutes, but if he happened to be there when he returned he would hand him over to the police. He returned, Brasillach was still there and Pisani handed him over to the judicial police. Brasillach was soon tried and executed for treason and complicity with the enemy.

Edgard Pisani also witnessed the agony of Pierre Laval who had absorbed a dose of cyanide and was present at his execution. He had followed Laval's trial which he judged harshly because the prosecutor tried to unravel the most intimate thoughts of the accused. For Pisani, this trial should not have been that of a man but of a politics. The young man imagined himself in the role of a prosecutor who pronounced his judgment in three sentences: "You have said: 'I want a German Victory'". "You have signed the texts that committed young Frenchmen to work-camps and to death". "Mr. President, History has judged: I call for the Pierre Laval to be sentenced to death". For him, at a certain level, history is the sole judge no matter what the intentions and motives of the perpetrators happen to be. Politics are cruel and demand that individuals be judged for their acts and their results and not for their intentions.

He was soon afterwards appointed Prefect of the Haute Loire, one of the one hundred administrative divisions of the French territory; then Prefect of the Haute Marne, where he encouraged and favored the modernization of the rural world and encountered firsthand the problems faced by the inhabitants of these agricultural regions. He learned patience; how to distinguish the simple from the complex, the relationship between the whole and the parts; the dialogue instant-duration that the political handles so badly. He learned the dialogue unity-diversity that should be the purview of the political; the complementarity disorder-order; the former creating and throwing out ideas, the latter executing them. And finally, the dramatic dialogue State-Society which distinguishes France a country where the unity was created from the centre, from England, Germany, the United States, Italy and even Spain, where nobles, founding fathers who, even though they were highly suspicious but could not do without it, tried to create a minimalist State.

These mutually inclusive and complementary dialogical entities were at the origin of Pisani's conception of politics. Since neither of the two terms was mutually exclusive, politics became the mediator between them. He progressed even further and attempted to understand the world in which he worked and the professions he exercised. He began to compare the politician to a general practitioner who deals with the whole organism and not with such and such an organ. Both are destined to decline because of the emergence of specialists: politicians are being replaced by economists and general practitioners by surgeons. For him, in establishing a cast, a corporation, a world, a specialist among other specialists the world of politics favored this decline, and the modern world is now suffering because of it. People today are beginning to despise politics because it isn't where they think it is; and they are either amused or impassioned by the spectacle it puts on.

He came to the conclusion that Politics must be reinvented, starting from a few simple ideas that are difficult for people to understand because they are simple. The economy has a tendency to impose its power on the world and, under these conditions, whether you like it or not, politics tends to become its servant. The disequilibrium created is all the more dangerous since it is founded on an apparently simple idea (prosperity), it makes use of science that it has subjugated, it also uses considerable means of "persuasion" and constitutes a real ideology that judges everything from the point of view of the market. The disequilibrium can be redressed or abolished only if it is opposed by a reality that is as powerful and as legitimate as it is. Such a reality exists in the form of society. It exists but it is not yet constituted. It exists against, but not in itself. By ensuring the mediation of different alternatives and by using every possible means to constitute society, the role of politics is to be the general mediator, and not an actor among others. This is what Pisani's long experience of agriculture, rural societies and the environment taught him. But in this scheme of things political power must be founded on mediation, and not the contrary.

These brief narratives encapsulating some of the salient moments in Edgard Pisani's life present a seamless continuity in the development of his philosophy and attitudes that continued to evolve over the last sixty or so years. However, I would like to highlight briefly a number of breaks he experienced that reoriented and gave new meaning to his existence.

The first break occurred, when as a young man he abandoned the Catholic religion. In his early years he thought about becoming a priest because of his religious family and school upbringing. Growing up as an adolescent he was a good student and began to question his Religion professors who rebuffed him in the name of authority. He began to doubt and, when he returned to Tunis after several years in Paris, he informed his father that he no longer believed as he considered himself to be a mystical agnostic. He evolved over the years as a person who respects all the beliefs and non-beliefs of those around him. The Church raised more problems for him than religion did and he considered belief to be a personal matter. Pisani practices a form of humanism that is based on the existence of people, values, ideas, places that must be defended and this leads him to distrust science, the economy, information, politics and the offhandedness that make some think that individuals can now do anything they want, that they can profane everything.

He fears that people can now claim all rights over everything, that they think that liberty consists in not recognizing limits, in not respecting anything. For him, one must first of all respect life and others, oneself and one's Country. These core beliefs informed much of his administrative and political activity.

A second series of breaks occurred when Edgard Pisani distanced himself from Prime Minster Georges Pompidou's government in the spring of 1968. When asked to become Minister of Agriculture in the summer of 1962 by Prime Minister Michel Debré and President Charles de Gaulle, he became a Member of Parliament without adhering to the Prime Minister's majority party. During this period, Pisani had a large number of bills passed by the Parliament and maintained cordial and positive relations with the majority and the other political parties. His independence, however, did create problems because, on the one hand the United Democratic Left in the Senate was a loose association but not a party and, on the other, he found it more and more difficult to agree with the policies of the majority and began to take his distances. His admiration for Charles de Gaulle remained because of all that he had learned from him; he was faithful to Michel Debré and held Georges Pompidou in esteem, but he distanced himself more and more from those who claimed to be their heirs. Finally, he wondered if, in diverging with the de Gaulle's policies and all the while wanting to succeed him, Pompidou didn't actually weaken the political architecture that he intended to inherit because he and his party privileged their own duration to the detriment of the duration of a certain idea of the Nation.

From the very first Pisani pushed for a strong European Union convinced that this was the only way Europe could become a world power. He thought of the Union as an ally of the United States but substantially different; as having privileged relations as a good neighbor of the Soviet Union; as becoming a powerful but unpretentious interlocutor with China and India, while being particularly attentive to problems of the Third World whose development it would encourage and help. He believes that this has not been the case since the Union has integrated a large number of countries that do not share its vision of the world and that have very different civilizations. It has not defined itself as a specific and autonomous entity and has an unsound constitution with no clearly defined international goals. It remains a non-power a non-identified political entity that is unable to assume the role of a power that can become the fifth member of the Directory of the World with the same trump cards as China, the United States, India and Russia. Finally, although the Union exists, it scarcely changes the way citizens live and exercise their political agency in their own countries. Although he is no longer a militant for the Union, Pisani stated he would vote for a European constitution all the while refusing the mediocrity it is floundering in. In short, he now refuses to despoil France of its political attributes to the benefit of an uncertain and powerless magma that privileges all that is world-wide and refuses to affirm itself politically.

Edgard Pisani joined the Socialist Party in 1974 and worked passionately to promote its success. He prepared a report on the reform of education in France that was not distributed because it introduced too many new changes. He campaigned long and hard with socialist candidates who taught him ways of thinking, living, working, of becoming a militant. He participated in section meetings, federations and twice in National Congresses, as well as in debates about actions that had to be implemented immediately and contrary motions where semantic analyses swept aside political analyses and even more important projects the militant-citizens needed. However, he had difficulty coming to grips with such a political apparatus; its endless public quarrels; the wars between leaders that always came to the forefront. Even though he did not renew his membership he remains faithful to its founding principals and votes for its candidates and their positions because in France today, for him the Socialist Party represents the Country's only chance one day to have a politics that mediates between the economic and society. The only hope the country has to build an authentic democracy where individual liberty and the common good are linked, where the instant and duration dialogue with one another.

The former Minister of Agriculture is currently planning a fourteenth book on the political where he will analyze the domain, activities, mission and responsibilities of all political formations but, interrogating the future he will attempt to define what politics should be today. In so doing, he will not begin with what exists now but with the needs that must be satisfied. Between Common Good, Power and Interplay, between Institutions and Civil Society, he will attempt to come to grips with what the Politic and politics should become because for him: "The former can do nothing without the latter, and the latter is worthless without the former".

In this latest book he plans to develop the ideas that govern and inform his beliefs we raised in this brief "Foreword". Although it uses politics, the economy remains separate from it and is about to dominate the world. As it is exercised and perceived, politics has become a game and elector citizens are disillusioned witnesses. Politics cannot be revived by itself. Only a new society can create the conditions for its revival by opposing its own dynamics to the dynamics of the economy. Hence, the future cannot be limited to social gains; it must occur within the context of the liberation and constitution of society as a dynamic, autonomous, demanding reality. If the political is the mediator between the economy and society, its power must be founded on its mediation and not its mediation on its power. Each person, therefore all citizens must not be content only to vote but must participate actively in the transformation and development of society. This will lead to democracy. Thus, society in its diversity and duration will fulfill itself and take up its share of responsibility.

Paul Perron
Adapted from notes and correspondence provided by Edgar Pisani.

Introduction

This book is not written by a farmer,[1] and even less by a peasant. It is written by a city-dweller who in France once was given the task of dealing with problems related to the domain of agriculture which has become marginalized in spite of the fact the world is totally dependent on it. This book has not been written to inform but to alert people, and to mobilize them. This includes agriculturists, but also those who are politically responsible, the economists, scientists, and citizens, since everyone is concerned. My aim is to make them react, as they are too often unaware of the very nature of the problem that agriculture has always presented, and continues to present to everyone in the world today. Not only does agriculture ensure our subsistence, but is impacts on our safety. It is the oldest of all political problems, and economists mislead us when they proclaim that this vision is no longer true, since we can get our food-supplies from world markets. Their discipline is not the only one responsible for issues of safety.

What is happening today causes me more worry, than it gives me hope. By wanting to "overextend" the land, we run the risk of exhausting it. By wanting to globalize the market, we deliberately ignore the fact that in their own way all the people of the world need to live from the labor of their land. By wanting to industrialize agricultural work, we end up driving out the peasants, and the cities and factories do not know what to do with them.

This book has six parts, and sets out eight proposals.
The first part deals with the discoveries made and the work undertaken during my "very long journey". Assessing a half-century of support for "productivism", something that distracted most of us from "the eternal order of the fields"; the second part is an invitation to search for a policy. The third part sets out in very precise terms a "European Agricultural, Food Producing, Rural, and Environmental Policy". The five "messages" in the fourth part, explain the different aspects of this new policy. It explains it to city dwellers, consumers and taxpayers; to members of the Agricultural Commission of the European Parliament; to ambassadors who, representing the Nations of the World, can compare hunger and waste, development and safety; to scientists who have

---

[1]In his glossary E. Pisani defines various terms that he uses that do not necessarily conform to dictionary definitions. For him, the word "cultivator" refers to a profession and is expressive and neutral. The word "farmer" retains only the function of production whereas the word "peasant" evokes an activity, a state, duration and roots. He notes that when he uses the word peasant he does so with respect, since those who are so named bring life to countries and country-sides.

contributed so much and whose excesses frighten and threaten us; to young farmers, finally, who must deal with the complexities of a profession that still exists today, but is no longer the one their elders practiced. A conclusion presents some thoughts on "Life, People, Land, Time and Politics". As an appendix, a "glossary" explains certain technical words and defines the meaning of other words that experience has lead me to assign somewhat different meanings from those given in dictionaries. And finally, a "Postscript" that discusses what happened to the book after in appeared in French in January 2004, and also the progress I have made since then.

Let's now turn to the eight proposals in question:

1) Agriculture, the food industry and trade ensure our subsistence, but a policy must guarantee the quality and quantity of the products concerned.

2) Even if it had the capacity, the Western World cannot realistically think it can ensure the subsistence of the nine billion people who will inhabit the earth in 2050. To do this the world needs all the agricultures of the planet. Each country has the right to feed itself. The poorest ones cannot exist otherwise. For the world to find its stability, a productive agriculture that respects nature must be established, at the same time as a modern form of agriculture that nourishes and sustains on the land the vast numbers of rural inhabitants the city cannot accommodate.

3) In the world, and in each and every country, the struggle against poverty and hunger does not exclude charity, but we are dealing with a matter of justice. However, this struggle should also be subject to the principle of prudence.

4) The Western World has a model of consumption and a standard of living that are wasteful, and if they were to be generalized, all the natural resources of the world would be exhausted in a single generation. By living without wasting, the Western World can live better than it does now. This is what it needs to invent in the future.

5) Since prices across the world are depressed and variable, both the European Union and the United States hand out agricultural subsidies that are often greater than half of their farm revenues. It is time for these producers to sell their produce in their own internal markets, at prices that actually correspond to their real costs. The countries of the European Union must do this in such a way, that by exporting their surpluses they do not compete with the agricultural products of developing countries.

6) Agriculture must respect nature. Farmers must maintain it. Production is no longer their only activity. They now must carry out tasks for the common good that affect the natural environment, and rural society. These tasks must be identified and recognized.

7) Medical research, together with agronomical research, has led to major progress in matters of health, diet and reproduction. They have reached the point, though, where they do, in fact, put into question our natural legacies and, often, our deepest convictions. It is neither the role of scientists, nor of the market, to rule on their legitimacy. This is for society to do, after having been educated by scientists.

8) Wisdom has it that global society and the agricultural world, in each national or regional political entity, must reach a pact to ensure that society has the safe food-supplies and the proper environment it needs, and that each political entity has lasting stability without which it cannot carry out its tasks.

Invitation to an Open Debate

Once I had articulated these eight proposals, I became aware that they had never been the object of public debate. It is simply by force of habit and tacit convention that people answer, or believe they have answered, these issues. As a matter of fact, none of the major issues we face today is dealt with in a way that triggers debate between all interested parties. The only debate taking place is between farmers and the government. This has never been sufficient, nor is it any longer acceptable because agricultural activity now creates problems that concern society and national territories, qualitatively safe food-supplies and environment. These are problems that farmers cannot solve by themselves. Even though they have the duty to decide, institutions do not have the capacity to express the real needs of society. If, in the past, society wished to express its needs, it certainly lacked the means to do so. But with the current electronic revolution taking place, each individual can have an opinion, and everyone can exchange ideas, above and beyond their social and political loyalties. The multi-faceted dimensions of the agricultural, food producing, environmental and rural problem strongly suggest that the information the political sector needs in order to deal with current and future problems must be found in society itself. Information is what is needed, for it is simply a question of informing better those who incontestably have the responsibility of making the necessary choices, and of defining the Common Good. The electronic revolution makes all of this possible. But it is also legitimate to ask what other route we could use to bring to fruition the "Pact" that Global Society and the Agricultural World need?

These are the reasons for this "invitation to a public debate" on the proposals and propositions presented and defended in this book. A WEB site is now open entitled "An Old Man and the Land"[1] that makes it possible for an exchange between participants to take place. It will enable me, theme by theme, then globally, to answer the criticisms that will be addressed, and the questions that will be asked. If these debates give rise to new ideas, an open group will be established to present new proposals, and new propositions. An organization then will be set up to obtain the best possible results from these exchanges.

---

[1]The WEB site (vieilhommeetlaterre.com) was set up by the French newspaper *le Monde* and ran for several months from January to March 2004.

Part I

Such a Long Journey!

Chapter I

First impressions, Child, Prefect and Senator

As a child, like everyone else, I ate fruit, meat, bread. I drank milk, without asking myself how all this arrived on our table; without asking myself how they all came into existence. And then one day, I saw some magnificent cherries swaying among the leaves of the great tree in our garden. A few weeks later, figs appeared on the branches of our fig tree! I saw a small donkey mount another! When I questioned my father, he said that everything that was on our table had been produced by those who worked the land or raised animals. He explained to me that his job was to extract minerals that existed in the ground, whereas the farmer's job was to milk the cows, and extract from the trees and the land what would not exist without his labor. He told me a whole lot of things that I have forgotten, but that still remain, nonetheless.

A few years later, with a few friends, I climbed the Bou Kornine, a mountain that symbolizes the Gulf of Tunis. While climbing back down, we see an old man lying in a ditch. He is breathing with difficulty, and moaning, but doing so, quietly! He is exhausted, and he smiles briefly, while one of us goes to fill his canteen to give him a drink. But the man does not have the strength to lift his head. With a single gesture, he asks us to leave him alone. He smiles once again, and says "Allah Agbar", God is great, then collapses, and dies. We are totally lost, and ask ourselves why this could happen. A small gathering of people gets together. A young man who speaks French steps out from the group, and we ask him why the old man died. "He has walked too much and he comes from the desert. This is too far." Each year in winter the tribe's people go south, and sow seeds of wheat that they cover over with sand. One out of three or four years, when it has rained, seeds germinate and prosper. The other years, there is nothing or almost nothing. As a few drops of rain fell in the spring, the tribe set out to go and harvest. This was all for naught; the tribe's people came back empty handed and continued their long trek north, with the hope of finding something to harvest. "Allah, help us!" Then they left, carrying on their shoulders the cadaver rolled up in rags. This memory haunted me. This calm and tranquil death was, more or less, the one that awaits the fellahs, the peasants. The history of our Middle-Ages has confirmed this: the glebe, serfs, and famines. This happened in 1930. Then, my life and my mind went off in other directions. It would be years before I had different impressions about agriculture.

During the Second World War, I undertook a number of expeditions to acquire some food from farms I had visited. As I was six foot four, ration coupons hardly sufficed to nourish someone my size, from one season to another. It was in 1943, in Entrammes and in Lassy, that I really discovered the land, the peasants, the farms, the groves; the enclosed fields against which, later on as Minister of Agriculture and under the pretext of regrouping the land, I would lead an excessive campaign. I stayed for some three months with Yves and Suzanne Le Carbont. He was a primary school teacher in a district where rural leases made it mandatory for farmers to send their children to private schools, and he had acquired a certain authority in the surrounding districts by giving advice and counsel to farmers. He "placed" me at La Pommeraie where, every morning except Sundays, I went just before breakfast to put in a full day. I was greeted with solemnity by Monsieur Jules Peslier, with a sly smirk by his older sons, with kindness and reserve by the mistress of the house, and I sat at the other end of the table with the hired help. Lively conversations took place about the village, the farm and work. For a few days, I kept totally silent. I learned to listen. I observed all of them, Madame Peslier in particular who did not sit down to eat with the others but served us, helped by her daughter. I was seated in front of the great fireplace, with its great andirons, its trammel, and the pot where the water sometimes sang. There was a big armchair and a small table, in front of the fireplace. One day, in the middle of the meal, the Master asked his youngest son to bring him his book. He opens it, and turning to me, asks me to explain the meaning of two words he had not understood. He was reading a history of rural France. This was the premeditated end of my novitiate. I answered the best I could. At the next meal, I was seated closer to the Master, who now often invited me to take part in conversations held around table. A few weeks later, I sat on his right in the eldest son's place. Our topics of conversation had changed. It became customary to question me about Paris, about my studies, about my native Tunisia, and about the war. I also asked them a few questions. We were conformists without really being so; free to speak our minds, curious about everything.

What would happen to me once the war and the occupation had ended? I wondered about this, but was quite incapable of giving an answer. I learned on the farm. In three months I carried out many tasks, more related to plants than to animals. Once summer arrived, I scythed. Then, it was time to harvest the beets, to pick the apples, harvest, participate in threshing grain, and the difficult task of building high haystacks. I even carried sixty kilo bags on my shoulders. I carried them up to the wheat loft, on a shaky ladder. All of this under the mocking gazes of the boys, and the worried look of their father. I was motivated by the pride of a city dweller who did not want these country bumpkins to show him up. I truly loved this initiation. But I had to leave. There was the celebratory meal on the appointed day, a Sunday. The primary school teacher and his wife are invited. I arrive at La Pommeraie ahead of them, just as Monsieur Peslier had asked me to. He has just returned from Mass, dressed in his Sunday best. He takes me by the arm and leads me around to see his various parcels of land, commenting on each and every one of them. We go to the stable, where he calls each animal by its name: where each day I saw Madame Peslier milk, and give the cows, as well as the calves, huge pails of water to quench their thirst. The farmer talks about renewing his lease, and also about his succession. He hopes that one of his sons will find a wife, and farm somewhere else for, "There is not room for two here". We visit everything, the garden, the chicken coop that is part of the "house", the domain of the women. Sometimes, he asks me for my thoughts. Before rejoining the others, he stops, standing firmly on his widespread legs. He is as moved, as I am. He tells me gravely:

—"You're going to leave…you're going to go far…you'll be either a Prefect or a Minister"

—"Why do you say so?"

—"A city dweller who can adapt here, as you have, can take up any profession. See, today, you speak about the same things we do, you speak as we do. You have taught us many things."

—"I've learned more in three months here than I've learned in ten years from my teachers and my books. Can I come and see you?"

—"You'll always be welcome here."

When, in my second posting, I was appointed Prefect of the Haute Marne, I returned to see him in 1947. He was the first person I visited. We chatted for a few hours. He wanted to tell me how he was worried about the future, about his children. Through reason and instinct, he already knew that many of the farms of Entrammes were slated to disappear, "because we need more land than we have". I learned that day, that farmers who live in isolation and read only a few books and newspapers have a pessimistic vision of the world. But all in all, they know so much: "The land teaches us more than all the books; because it resists".

In 1945, for my first posting I was appointed Prefect in the Haute Loire, a Department that represents the broad mass of French people, away from all major routes and protected from all invasions. Only the Pilgrims of Saint James of Compostella have come through this mountainous and poor region. When I arrive in the city of Puy, I learn that no regrouping of lands[1] has yet been envisioned, that the farms often have

---

[1]Beginning in 1907 France instituted a policy of reconstituting agricultural domains that had been excessively parceled into small plots making farming very difficult. Lands were regrouped through exchanges and redistribution in order to reconstitute more viable and more workable farms.

a communal room with a floor of packed earth, hundreds and hundreds of farms are isolated, it is often necessary to go and fetch water for the animals over a long distance. Having learned that there is a country fair here and there, every day of the week, I decide to visit the markets. This turns out to be a surprising experience. The people are all dressed in black, wear the same round hats, the same boots with rounded toes, have the same unexpressive faces. The men have walrus mustaches, and they speak little. They have come here to sell a calf or a pig; to buy tools. Their wives accompany them, selling poultry or vegetables, and buy clothing. They pretend they don't see me. In certain districts, the Prefect is the one who, in 1905, oversaw the "inventories" of Church property. Here and there, a young man in a plaid jacket, head bare, wearing brown boots, comes up to me and says awkwardly that "the land no longer feeds its people". Bolder that the others, a forty-year-old man and his spouse say seriously, "we have four boys. We're not certain a single one will stay on the farm. We decided to send them to school. But it's very expensive. They will only continue if they study hard".

In 1946, we are still in a period of post-war rationing. Considering that the Department[2] is self-sufficient, Paris decided not to supply it with wheat. We eat poor quality bread made of corn flower. I decide to take a trip around the districts to meet the mayors, and the "important" farmers. I give my speech on the sorry state of the countryside, and on the even more sorry state of the cities. I launch an appeal for solidarity. Near Yssingeaux, with everyone feeling good, a man who looked like a Gallic leader who defended the country against the Romans, emerges quickly out of the crowd, and tells me resolutely and solemnly: "Times are hard but I certainly would give a few sacs of wheat for Old Pétain".[3] The Deputy-Prefect shudders. I calm him down, because I prefer a few sacs of wheat to an act of vengeance. I leave with the promise of two or three tons of wheat, and the state of mind of a watchmaker who has to set thousands of watches on time. People here do not believe that progress is possible; for them the future will be like the past. It is not despair. I do not feel any revolt in them, but a pathetic resignation. Thirty years later, I found the same thing in Africa. The elected officials tell me to make no promises, so as not to create expectations. I am

[2] One of 95 administrative divisions of France placed under the authority of a commissioner of the State (Prefect) who is assisted by a Council General.

[3] French military man and politician (1856-1951) commander in chief of the army in WWI was promoted Marshal of France in 1918. He became President of the Council in May 1940 and signed the armistice with Germany the 16th of June 1940. After the war he was condemned to death for collaboration with Germany. He was subsequently pardoned and served the rest of his life in prison.

revolted, and begin to work on the first "Departmental Plan of Modernization and Public Works". I discover the cruel alternative that France experienced after the war: Must the state and its elected officials go along with this resignation and treat only the general discomfort that results from this, or must they sound a call to action and "mobilize" men, women and adolescents against resignation and the decline that accompanies it. I chose to be a live wire. Why wait until it falls apart? One has to create a movement out of this despair. One has to foster revolutions that cannot be avoided.

Two years later, in 1947, I moved on to Haute Marne between the Champagne and Burgundy regions of France. The valley of the Marne, the Plateau of Langres, have been invaded over the years, they experience waves of tourists going by; for centuries they have been the home of numerous industries, especially metallurgical. Workers are conscientious, and so are the peasants, whose guiding light is Eugène Villemin, the President of the National Federation of Farm Unions (NFFU), a single and solitary unsociable shepherd, a formidable orator, extremely well read, a courageous free thinker, and an unrepentant hunter. He could have been my adversary, but he became my very demanding partner. Industry is lead by President Godinot, a wise conservative. The world of the workers is defended by various union leaders and by an elected member of the communist party, Marius Cartier. As we are in the eastern part of France, tensions and conflicts do not put into question mutual consideration, and the administration is respected. Departmental heads are quality individuals. Among my three divisional heads, the first is a worker who works, I have faith in him, he meets all my expectations but he annoys me; the second is a lazy body who hardly works; what can I do? The third head is a lazy body who works: he is a wonder to behold. I have used this method of classifying my collaborators, all my life.

I focus on four problems: The construction of Saint Dizier le Neuf that I will not talk about here; the development of the cooperative agricultural movement; rural development; and the forests. As I proceed with my plans, I will discover a number of things that accompany me throughout the exercise. In a bold and quiet way, the cooperative movement expressed the revolt of farmers against the major situation facing them, and against the difference between farm prices and consumer prices. It expressed the panic-desire that men who, isolated in their work, had to band together and become a real force. The revolt movement stopped there. After cooperatives were established, the general assemblies were gloomy and controlled by administrative, commercial and technical directors who took power after graduating from school. The members of the cooperatives became strangers in their own homes. As they developed, the groupings of cooperatives took on the tone of capitalist conglomerates, difficult to control. I saw the incredible development of the cooperative cheese cellars of Langres; I heard the members express their interests, and soon noted that they had become insipid. The results were positive, but the enterprise was too big for a stockbreeder; he counted in tens of thousands of francs,[4] whereas he was presented with accounts in the tens of millions. Another phenomenon cropped up: to amortize its enormous investments, a cooperative deducts money for each litre of milk it processes. This investment that increases production cannot, statutorily, become part of the "contributor's" patrimony: the cooperative is, indeed, a corporation of individuals; its shares are non-negotiable, it is not a limited company, where each holder owns stocks. In spite of its legitimate constraints, the privileged tax policy enjoyed by cooperatives has progressively been eroded. Today, I blame myself, as a former minister, for not having paid sufficient attention to this problem. I blame myself, all the more so, since this post-war cooperative adventure has been a good one, in particular the Crédit Agricole that has now become one of the leaders of the French banking system.

I especially resorted to this bank to bring about or, rather, to facilitate the rural development to which I dedicated a good deal of my time. Among the five hundred similar communes, Bassigny comes to mind, where water conveyance, sewer networks, sidewalk curbstones, planning, where to put piles of manure were all implemented! It all began one evening, at the end of a meeting when the Mayor and I met a very outspoken matronly woman. She said off the cuff: "Mayor! Schools, roads, are of no use. There will soon be no more women to marry in the commune. This is not a life. Milking cows on Sunday evening, instead of going dancing; having to wear boots any time you go out, because of the liquid manure; leading the animals to the drinking trough twice a day; breaking the ice in the fountain in wintertime…Young women have had enough". We answered with pious words; then, we decided to meet at the Prefecture. We worked out a multiyear program, to meet the woman's requests. After several meetings over a period of a few months, we agreed on a plan. But we needed resources. Now, to make people want water conveyance that is both necessary and profitable, the Minister of Agriculture wanted cheap water. On my advice, the General Council of the Department sets in place an accelerated plan of water conveyance that asks for more funds from the consumer than from the Public Treasurer. The Department of the Haute Marne is modernized in very short order. In the meantime, I become a candidate for the Senate

---

[4]The "Old" Franc was devalued in the early seventies by 100 with the advent of the New Franc that was worth roughly 20 cents at the time. Ten thousand francs in the 60s were worth some $20 U.S. before devaluation.

and campaign on modernization, and what it would cost. After my election, I no longer believe in the stupidity of the electors, but in the stupidity of those who believe electors are stupid. France is a reasonable country, all one needs to do is too explain things, and set out objectives.

The Haute Marne, also teaches me a love for the forest. I always took pleasure in walking under tall trees, sitting on a tree trunk listening to the silence broken by the murmur of leaves, the song of birds and, sometimes, the gallop of a stag and a deer. But that is where I discovered the forest as a living and immutable being, which is also fragile. It is in the forest where I experienced time and understood that it does not belong to us, and we are only its beneficiaries. That is where I experienced one of my greatest fits of anger, and committed one of my boldest acts. One day, when I was invited and guided by Michel Cointat, a competent and passionate forester who would become the director of my cabinet, then Minister of Agriculture, I visited a patch of forest that had just been clear-logged. Nothing was left standing. What a disaster, and what a ruin: "it will take a century to reconstitute what existed yesterday. Having reimbursed himself by selling the logged trees, the owner will then sell the land, and it will all be clear profit for him". He later informs me that the forest of Arc en Barrois, ten to twelve thousand hectares, will be put up for sale by relatives of the Comte de Paris[5] who own it. I ask to meet the Comte, and he grants me an interview. He lets me know that the forest belongs to his relatives, and that he can only draw their attention to the serious consequences of a sale that does not guarantee it long-term viability. In case of an offer the administration would be informed, the Department then can become the buyer with the help of National Forestry Funds. The Haute Marne now owns the property. Having learned of my meeting with the Comte, Jules Moch, Minister of the Interior, summons me and gives me a thorough talking-to. But in conclusion, he tells me I was right to do what I had done. I had won. I love the forest more than I do trees; each one must disappear, and each one must last; much in the same way a human community must. Each tree lasts fifty, one hundred, three hundred years, the small ones growing in the shade of the larger ones, each one nourishing itself from the humus that the leaves renew every year. It is when walking in the forest that I made a significant comparison: "humidity, humus, humanity, humility". This remains like a Commandment with me. My passion, and the sadness about the hundreds of thousands of hectares abandoned to juniper bushes in my country, led me to create, then to preside over the Society for the Development of the Fallow Lands of Eastern France.

---

[5] Pretender to the throne of France, son of Jean d'Orleans, the Duke of Guise and of Isabelle Orleans both of whom are descendents of Louis Philippe, King of France.

What had to happen did. Having spent six years in Haute Marne, I knew that I would have to leave for another Department. I said so in those very terms at the banquet of the great animal fair of Montigny le Roy. Two Senators are present. Eugène Villemin tells me that I must stay, since the Department needs me for a few more years. I tell him that six years I spent there, are a record. He interrupts me, and says off the cuff, "run next year for the senatorial elections. You'll be elected". This is the first time I ever considered running for office. I consult with my close collaborators. I resign and move to Paris where, three months later, I learn that Senator Barré died during the night, while driving his car. The election will take place in two months, whereas the minimum legal period of non-occupancy is six months. The Democratic Left decides that I am their candidate. I am elected and my election is not contested by the Senate. I replace Barré on the defense and construction committees. I only have the weekends and holidays to keep in touch with agriculture. I continue my interests in the forest, rural development, much more than in agricultural production.

As a delegate to the Council of Europe, I am responsible for writing reports on problems of energy! I continue to do my work in Paris just as in Strasbourg, but I am bored by it all. The job of Prefect has given me bad habits. In 1960, I decide to leave politics and to look for work elsewhere. In May 1961 I tell Michel Debré[6] that I am going to resign. He does not believe me. I am about to do so when, on July 3, the Group of the Democratic Left asks me to follow from beginning to end the agriculture debate that is about to unfold in the Senate. Fifty speakers are scheduled to speak. I listen to them, and they all say the same thing: things are not going well, the State must pay. At midnight, there are still twelve or fifteen senators present, and I am the only one not signed up to speak. President Monnerville[7] is in the chair. I go up to have a few words with him, and to inform him about my weariness, and fears too; as all this rhetoric does not seem to lead anywhere. "Do you feel like speaking?"—"Probably not."—"Go ahead". He signs me up on the list of speakers with his big blue crayon. I am the last to speak. I run back to my bench. Having quickly decided not to repeat the same speech I

[6]Michel Debré (1912-1996). Mobilized in 1939, he was a militant in the French Resistance. He was an ardent defender of "French Algeria", and played an important role in the recall of General de Gaulle in 1958. He was Prime Minister from 1959 to 1962, when he was replaced by Georges Pompidou. He also was Economic and Finance Minister, Minister of Foreign Affairs and then of National Defense during the latter half of the 1960s.

[7]Gaston Monnerville (1897-1991) was Member of Council in 1946, then President (1947-58). He was then President of the Senate (1958-68), while he was in opposition to the Gaullist regime.

heard fifty times over, I decide to look at the agricultural problem from a city dweller's perspective. I scribble some notes. The President soon calls me up, since at this early morning hour every one is brief, and only hopes that his local newspaper will report on his short intervention. The Minister listens to me; he will answer tomorrow afternoon; the session is ended.

He answers in the presence of the Prime Minister, and spends two thirds of his intervention on the remarks I had made. In the grand salon of the Luxembourg,[8] I meet Michel Debré who tells me "Edgard, don't leave!" I do not answer. Rereading my remarks recently, I found them quite daring: "…the State gives bottles of oxygen, and responds instantly to appeals…the State is ready to give money to maintain the misery of the agricultural world; it is not ready to give money to create new conditions for exercising a profession that will renew itself…Parliament has serious responsibilities for this state of affairs. For the longest time it has considered the agricultural world as a political mass that has to be satisfied, and not as an economic entity that has to be organized…Subsidy and protection of the humus gives agriculture a character of quasi- public service…In truth, what is profitability? Is it an accountant's notion, in which many of the activities we hold dear would not be profitable, and would have to disappear. Is it an economic notion, in which a certain number of activities which, from an accounting point of view, must disappear, could be saved… To be fruitful, the dialogue "nation-agriculture" must be harsh …Agriculture has created for you, has created for us, many problems. It deserves to be answered with the very same rigor, the very same firmness and, for its own good…I am one of those who do not defend the little family farm, for there is one word too many in that expression, and that word is "little". We must not defend the farm because it is little, but because it is family…If it is true that no reasonable economic means can help a farm attain an economic balance, we have to accept that it should disappear. We do not have the right, in the name of the monstrous respect we hold for what is little, to maintain families in slavery on farms that can never feed them. It is our duty to affirm that we are seeking that fine balance between the family that cultivates, and the land that must feed the family…"

By offering me the position of Minister of Agriculture, the President of the Republic, Charles de Gaulle, and Prime Minister, Michel Debré, were therefore endorsing a program. I remain surprised by the audacity that my agricultural inexperience gave me; this is where I find the explanation for my passion for interdisciplinarity, my suspicion of anything that encloses one in a singular vision of things. This explains my suspicion of the vision that the agricultural leaders revel in Much later

---

[8] The Palace of Luxemburg in Paris is where the Senate meets.

on, I would have the irritating experience of this in the Social and Economical Council of Europe where, in 1994, as the member of the agricultural section who was responsible for giving a report on "French Agriculture and the Common Agricultural Policy" I saw my text prevented from being presented in a plenary session by representatives of professional organizations, in spite of having been unanimously endorsed by all those who were not farmers. The report presented the agricultural problem as one that concerned the entire French population, and not only those belonging to a single profession. This is criminal! The Economic and Social Council would be much more daring if each of its sections was not the private domain of the profession or the professional category whose problems it studies.

Chapter II

The Ministry of Agriculture (1961-1966)

A few days after my daring intervention of July 3, we leave for a family vacation, and go down to the south of Italy. We return on the 21st of August since I was responsible for a report on energy and had a meeting with civil servants from the Council of Europe on the very next day. We had hardly begun to work together when the telephone rang.

— "Mister Edgard Pisani?"...I will put you in touch with the Prime Minister"

"Edgard?"

—"Yes, good day Michel"

—"Do you want to be Minister of Agriculture?"

—"Hmm"

—"This is the only effect this has on you?"

—"The farmers are taking to the streets...I will take the train and be there tomorrow at your office, whenever you want"

—"The General wants to sign the decree tonight, before leaving on holidays. Can I call you in an hour?"

Surprised, and...overwhelmed, I collapse in an armchair, before getting up and calling Pierre Pflimlin,[1] who cannot be reached; Edgar Faure[2] tells me that Valéry Giscard d'Estaign[3] is a candidate; he advises me to accept; Paul Driant, Senator from Moselle was approached this morning. He refused since as a professional national leader he cannot accept this appointment, because his colleagues would consider this to be a sort of betrayal. Eugène Villemin, my favorite interlocutor from the Haute Marne, urges me to accept. The same as Driant, he tells me that the agricultural leaders will react positively to my candidacy, as they appreciated the logic of my speech. The telephone rings, it is Michel Debré.

—"And so?"

---

[1] French politician (1907-2000) who was President of the Council during the Algerian crisis of 1958 and Minister of State under Charles de Gaulle. He also presided over the Council of Europe from 1963 to 1966 and the Assembly of European Communities (1984-87).

[2] French politician (1908-88) who was head of several ministries during the Fourth Republic (1946-58). He joined the rank of General de Gaulle in 1963 and was Minister of National Education after the May Revolt of 1968.

[3] French politician (1926—). Minister of Finance 1962-66, he succeeded Georges Pompidou as President of the French Republic in 1974, and was defeated by François Mitterrand during the next presidential elections in 1981.

—"I accept but on three conditions: I alone will name my cabinet; it is the Minister's responsibility and not Matignon's[4] to set agricultural policy, and it is by me that agricultural policy must be presented to the government; finally you will make available a gift of joyous Coming that will permit me..."

—"No! Our agricultural generosity is ruining our budget."

—"Let me, at least, when the right time comes, talk to you about my ideas"

—"Hmmm...Yes. And?"

—"I agree, thank you"

—"I will call you back in a half hour; the General wants to see you tomorrow, at Colombey. I will see you as soon as you arrive in Paris, tomorrow."

Here I was hoist on my own petard, and responsible for putting into practice the program that in all innocence and on the eve of quitting politics I had sketched in an improvised speech. The Prime Minister and the professional leaders know what I think. Let's get on with it.

On August 22, at 11 a.m., I finally discover La Boisserie, where General de Gaulle lives. I am not afraid of Him, but I am afraid of facing Him. He greets me at the door of his house, leads me to his office, with windows overlooking the forest. I remember a single sentence from this meeting, which mainly focused on Europe: "You are not the Minister of the farmers but the Minister of Agriculture of France." This aphorism would haunt me for six years. I looked for, just as I am still looking for, the right balance between common agricultural good, and common ... common good. We have to find this balance. I set this as my objective, in the car that was driving me to my office which will be my "home" for six years. Did I succeed?

When I get to rue de Varennes,[5] I call Matignon; the Prime Minister will meet me at the end of the day. I ask all the directors of the Ministry to come and chat. I meet with them all together. We have a long conversation, the first of a series of weekly meetings where all problems are discussed together, openly and among equals! A few words are exchanged on the crisis that is not over and on how my nomination will be received generally. In fact, the professional leaders know that they will not win against the government. They are afraid that they will not carry the rank and file, and that the movement will fade away; they hope that I will offer them an escape hatch. Two problems were identified after three hours spent conversing, thinking out loud: European policy that we will begin negotiating in two weeks time (the 5th of September), and the need to bring about a fundamental change in agri-

---

[4]Matignon is the Office of the Prime Minister.

[5]Office of the Minister of Agriculture in Paris.

cultural society that is itself divided about what needs to be done. The problem we were faced with became clear when I was walking over to Matignon with Charles David, the Director General of Rural Engineering. How can we create a European dynamics that brings about a transformation of the agricultural world, a revolution that satisfies it, while contributing to the modernization of France? Charles David said to me: "the leaders will make demands. You must be ready to make proposals. But no details, that is the role of the civil servants, us. You must give them a vision of a future that is different. Without doing so, we will all become prisoners of low-level discussions. Open the doors and we will occupy the terrain." In this way, the relationship between the politic and the administrative was clearly defined. This is how I understand the expectations of the farmers: as I said in the Senate, two months earlier, they have the feeling that we continue to give them "bones to gnaw on." This leads only to anguish and revolt.

Michel Debré warns me about the logic of permanent "compromise" that for political reasons, in France makes the government almost the only institution responsible for the future of national agriculture. From this very first meeting, I feel a deep respect for my interlocutor that will later become true admiration. He belonged to the species of top-ranking civil servants who are transfigured by an acutely developed political conscience. Grumpy, undoubtedly, sometimes a Bogeyman, but so generous, so bold, so courageous; what an open mind, no matter what some might say! Whether he agreed or not, it was a pleasure to work with him because he listened to you, left you free or decided decisively, but he was always supportive.

I got down to work. Rather than giving a day by day account of my term as minister, the longest in the history of Agriculture (it helps to have time on one's side!), I will take up some of the major files and, except for conjuring up a few anecdotes, I will try and set out what they taught me.

Brussels and Common Market Agricultural Policy

Those who do not have a clear idea of Brussels at its beginning cannot understand the dynamics of Europe. I do not know what feelings Jean Monnet,[6] Robert Schuman[7] and Edgar Faure experienced, but I will

---

[6]French economist and politician (1888-1979) who initiated the European Union and wrote the declaration of May 9, 1950 that formed the basis for the European Community agreement on coal and steel, better known under the name of the Schuman Plan.

[7]French politician (1886-1963), Minister of Foreign Affairs (1948-53) who implemented the Marshall Plan and reconciliated France with the Federal Republic of Germany. Along with Jean Monnet, he was one of the strong proponents of the European Union.

never forget the time when, on September 15, 1961, six European ministers of agriculture began to cross swords with one another, and defend the national interests for which they were responsible, but with the ambition of creating something new and original. It was not simply a case of reconciling Germany; nor only of constructing a Europe which would no longer wage wars. It was basically a question of constructing a new world, where through common interest nations that were formerly irremissibly hostile to one another would attempt to establish common policies that would make any conflict unimaginable. Because of this goal, we became dedicated and open negotiators, rational and bold adventurers. Above and beyond our own project, we thought that the world had eyes on us, and that another philosophy; other political institutions could come into being. All six of us sometimes paid a national price necessary to create this model, this future. Maurice Couve de Murville[8] felt this strongly when he told me one day: "my dear colleague, our objective is not for our negotiations to succeed, but for our interests to be safeguarded, at least as much if not more so than those of the five other countries." I then understood, and would understand more and more, the difference between diplomacy and politics: the former is the confrontation between contradictory interests that attempts to find a mutually acceptable code of behavior; whereas the latter seeks a common life founded on a common ambition that respects the balance of immediate and contradictory interests, with time leads to the creation of a Common Good. Couve de Murville thought that we should make the fewest concessions possible. My tendency was to think that a good understanding of our common interests would be more profitable than the stubborn defense of our traditional agricultural positions. We were both right and, when subjects under discussion and moments demanded, either his or my visions would reasonably carry the day. If I were to appraise the situation now; I think today that it is the cooperative interplay between both of these philosophies that enabled France to exercise the influence it did in establishing the Common Agricultural Policy (CAP).

I had the privilege of having two special interlocutors during these negotiations: Charles de Gaulle, the President of the French Republic and Sicco Mansholt, the European Commissioner in charge of Agriculture. I had numerous meetings with the General, and I even wrote an entire book on this. He sometimes met with me three times a month, as the subject both interested and worried him. He expected the negotiations to respond to our interests, but also that they clearly state

[8]French diplomat and politician (1907-1999) who was Ambassador of France in Washington and Bonn, as well as Minister of Foreign Affairs from 1958 to 1968. He was Prime Minister from 1968 until de Gaulle left power in 1969.

what the interests of our partners were. I sometimes had the impression that he was ready to consider the successes with respect to our interests as a series of failures confirming his pessimistic vision of the world. But all in all, I am certain that when we reached a positive conclusion on January 14, 1962, he accepted this with a certain amount of relief. Even much more than I imagined, when we reached a positive conclusion on January 14, 1962. He followed the debates in Brussels, and wanted me inform him or take orders before and after each European ministerial meeting. He had a way of proceeding that I still admire today: our exchanges began with the briefest possible account I gave of the situation and the positions (of the professional organizations in France, Germany, Holland, of the Commission.), as well as my own position. Then the time came for questions. He noted the account I gave, as well as my answers. Then, sitting upright in his armchair, with both hands flat on his desk, he would firmly say to me "we will do this or…that". This was not an order, hardly giving instructions; it was simply the definition of France's policies. One day, I had to inform him of a failure I experienced after he had said to me that "we" had to adopt a position that I had presented as being risky. He asked me to explain the reasons for my failure. We discussed them. "And now what?" I proposed that we maintain the same positions. "We will therefore…" It worked. Not once did he blame me for the failure in question. How could he have, after he had said "we"? The failure was his as much as mine. In Brussels, the feeling I had of always of expressing my Government's position, and the certainty I had of never being disowned, gave me a strong position that none of my other colleagues had, except for the Dutch.

You need several trump cards to spearhead international negotiations: expert and vigilant advisors, knowledge of the files, an exact idea of the objectives, alternative solutions to propose, but also political toughness. It is not sufficient to have courteous relations with your interlocutors; as this is the most mundane aspect of diplomacy. You have to reach a certain level of mutual confidence; a certain complicity. On numerous occasions I did meet face to face with each of my German, Belgian, Italian or Dutch colleagues, and came to an agreement on a point of contention. But together we continued to cross swords, until at a given moment we resolved the drama about to unfold by putting "the" alternative solution on the table.

The results obtained in Brussels were never criticized; yet, I did ask myself if my behavior respected the unwritten laws of French diplomacy. We are willingly doctrinaire rationalists; we have been a major power, which we are no longer; we are cruelly Jacobinists,[9] when a less struc-

---

[9]A person who is an intransigent supporter of the French Republic and a partisan of a centralized state.

tured vision of national and international institutions is imposed on us;
we are somewhat pretentious, and we readily become intolerable with
our interlocutors who unite against us; we are more attached to our own
interests than to common interests; in spite of the advantages those can
have for us; we have the ambition of triumphing over so and so when
we should be trying to overcome the difficulty together. We like to give
lessons. We are less appreciated than we believe. We are feared, we are
undoubtedly respected; others do not like to negotiate with us. Our
strength lies elsewhere. It is in the alliance we could bring about between
a form of Cartesianism[10] that enables us to clarify a negotiation, and a
form of inventiveness that we have inherited from our long history. I
give credit for this Cartesianism and this inventiveness to the teams of
civil servants that surrounded, and often guided me.

After the CAP was adopted, my professional interlocutors, my collaborators and I, attempted to determine the critical points in the plan
we had accepted. We retained four of them:—the fact that in spite of our
efforts all the produce, and therefore all the producers did not enjoy the
same level of support,—the fact that, for the produce in question, support was directly proportional to the volume produced by each farm,
which advantaged the biggest ones,—the fact that cattle feed was not
protected at any of the borders, while it would soon compete against our
cereals,—and, finally, the fact that these policies only affected produce,
and not the units and systems of production. Hence, the measures taken
did not respect the Treaty of Rome[11] for, even though they enabled
France to compensate for the risk it had taken in industrial matters, they
did not favor the family farm the way the Treaty did. But at no moment
did any of the participants even hint that a secret strategy had imposed
its laws on us. I realized this when writing this book; I discovered that,
even before the beginning of negotiations everything was rigged so that
the European Community would become the United States' agro-food
competitor. For the longest time, I believed that this result had been
obtained as a by-product. It fact, it had been sought from the beginning.
If I had not been so naïve, I would have interpreted the debate that
opposed the tenants of "prices" against those of "structures", Blondelle
and Debatisse, NFFU and National Centre for Young Farmers (NCYF),
in a very different light. In fact, Blondelle was one of the instigators and

---

[10]Someone who ascribes to the philosophy of René Descartes or his successors:
clear, logical, methodical, rational.

[11]The treaty establishing the European Community, signed in Rome on March
25, 1957. Its 248 articles determined: the founding principles; the citizenship
of the union; the community policies (including agriculture); the association
of overseas countries and territories; the institutions of the community; and
general and final provisions.

accomplices of the choice we made. This debate actually still reflects two visions of our agricultural future. Already in 1961-62, it foreshadowed the crises that would later rock the CAP.

But this vision was as good as it could be, considering the historical conditions that gave rise to it; it actually led to remarkable results. It gave a dynamics and an orientation to modernization. What would I have done with the leaders of the agricultural world if I had not been in the process of negotiating a future, creating the conditions that would ensure its success? I would have had to tell so many stories, to make the agricultural world understand! These stories would all be about the great power that the French agricultural world was given by Europe of the future. What would French agriculture be today, if it had not become capable of exporting to Europe and to the world? We needed a Utopia, but first of all a market.

"The Farmers' Discontent"

Agricultural leaders played a dominant role in this adventure, and they played this role as responsible players. Whether or not they happened to be conservative; these leaders considered negotiation and politics as an opportunity for the renewal of Europe. They were always present alongside the negotiators to express their expectations, to inform them, and warn them of possible errors. They mobilized militants and agricultural opinion in its entirety; they "mobilized" senators and deputies, also journalists, in such a way that, for the first time in French history, we had the impression of common interests, and a real solidarity between agriculture and society. This was obvious, and Georges Villiers, President of the Employers of France, one day complained to Georges Pompidou about the weakness of the Ministry of Industry. Teasing me, the Prime Minister asked me if I wanted to be responsible for "that too?"…No debates in Brussels, or government consultations in Paris, were prepared without the professional leaders. My interlocutors knew that I could not tell them everything, and understood when I asked them to help me to adopt such and such a position. I always got their support. It is with their help that I led off in my negotiations, and that I wrote most of the texts that I would then present to the Government or to Parliament. A pact existed between us that was never broken; I gave my word to take their position into account but I did not give my word to maintain it, for I had to take into consideration the needs that were not only nor always theirs. I never called that co-management.

It would be false if I were to create the impression that my relations were the same with René Blondelle and Michel Debatisse, with the NFFU and the NCYF, with the large farms in the north of France, and the small and medium sized farms crisscrossed by hedges and trees, with Brittany, the mountains, and the South. Some demanded that their situation be consolidated, others, that their future be assured. And still others would say loud and strong that the regions, the wheat and beet farms did not represent all of French agriculture. The former, would warn me against the temptation of favoring structures to the detriment of production. Each brought his own agenda to the table, but the necessary balance was soon found. Yet, the greatest joy I have of my time as Minister of Agriculture comes from my relations with the Young Farmers, almost all of whom belonged to the Christian Youth Agriculture movement. To describe these relations, I would like to talk about our "evenings". As soon as I came to the ministry, I tried to organize an evening of discussion every two months, to which I invited farmers, civil servants, teachers, industrialists. These free-wheeling meetings taught us a great deal: over the long, rather than the short-term. But there were urgent matters to deal with too, tactics to be countered, obstacles to overcome or make disappear, ideas to find, policies to be formulated and carried out. The only way to do this, was by meeting in a small group, by having secret discussions where one could say everything, without a "concession" made by one of the ministers or by one of his interlocutors being alluded to on the outside. This is true freedom of thought and discussion. Only the Young Agriculturists were ready for this, and were close enough to me so that I could take the chance of saying that on such and such a point I found they were undoubtedly correct, but I could not meet their demands. In this way, we made some surprising intellectual concessions that were never revealed. For example, ideas that were unacceptable one day; reappeared several weeks or months later and, as they became familiar, were soon accepted. This worked in both directions.

These meetings were held generally at my place of residence, Rue Bayard. A light buffet was set up to help us chat and exchange ideas, also to warn each other and sometimes say negative things. This took place over a three or four-hour period. We parted exhausted but happy. I soon learned that, after having left me sometimes after midnight, they continued on the sidewalk or had a beer together to be certain they understood the same things, and to get ready to move further ahead. This is how we invented the Social Action Funds for the Development of Agricultural Facilities (SAFDAF). This was not the result of an idea expressed by a single individual but came about during our evening meetings: at the beginning we thought of some measures to be taken, then the certainty that we had to put together a coherent set of policies with precise objectives to give these ideas a real importance. Still, these measures that were of a social nature, should in no way be considered as open to all, they had nothing to do with "social coverage"; as they were selective, they were to be proposed only if they had a structural impact. This would all take place in just one evening. The next morning, I met with civil servants in my office, and informed them of our deliberations. They began to draft; that very evening I presented the child to our older leaders; two or three days later, the text was discussed at the governmental level, and soon adopted. I have never ceased to regret that most of the major texts, and major decisions, are worked out by councils, or committees of specialists, who write before discussing, and who are then ferociously faithful to what they write. Time for creative liberty hardly exists in official processes, where each person adopts a position and takes public opinion into account before discussing with his or her partners!

Chapter III

Mission Accomplished

I therefore had to encourage a necessary transformation of agriculture in France, and follow it closely. It turned out at one and the same time to be harsh and complex.

In 1962, we had noted rather than understood the contradiction that existed between the European rules and production rates, and the social and cultural realities of the rural world; the contradiction that existed and would deepen between the different regions of France and, still more so, between the different regions of the Common Market. I had not understood to what degree there was a substantial contradiction between the "laws governing agriculture", and the European policies that were being put in place. Yet, all I had to do was go by helicopter from the Champagne, to Brittany, then from Northern France to the Limousin and to the Pyrenees, to realize that the same message was received and understood differently from one region of the country to another. But as mixed farming was very widespread, a great many farms benefited from grain subsidies. That is why they were all favorable to decisions made in Brussels. However, little by little the differences became known and protests arose from those whose food-producing, egg, or other markets were neither organized, nor protected by European agricultural policy. This is what motivated the increasingly vigorous debate between those who thought in terms of the market, and those who thought in terms of structure.

The NCYF became stronger, its ideas took shape. Blondelle and Debatisse took up the struggle, all the while trying to find a compromise. I was in favor of unity that took differences into consideration. The Ministry was dealing with a new generation that emerged from regions that had until now been somewhat unobtrusive. The great landowners, the classical heirs, the farmers with huge farms were located near Paris, and it was easy for them to lay siege to the offices of the Ministry, and to talk about "prices". But now, young men who had been trained in school, were used to debating, dressed any old way, sometimes ignoring the very rules of naïve and forthright civility, young women too, who came and spoke about their social conditions, the constraints of the garden, the henhouse, the stable, the spring where twice a day they had to lead the animals, about the medieval state of their farms, the regrouping of lands, water conveyance, the lack of telephones and, the education of their children. This was a real revelation, but these differences were perceived as a threat to the established order. How were we to proceed?

I never trivialized these contradictions, and I never let a difference expressed by someone go by without taking it into consideration. As for the essentials, I had to concentrate on unity, which was my strong suit when facing Parliament, dealing with the Government, with Brussels, with urban public opinion that was preoccupied with the "political" progress in the agricultural world, and the "privileges" that were granted to the "peasants". Unity was essential here, because one of my strengths, when dealing with "my own people", stemmed from the fact that I was their interpreter, and stood up for them everywhere, all together. Anyone coming from the outside in charge of carrying out reform can neither put into question the "essential" unity of the constituents making up his or her domain, nor suggest that he or she will reform, when the best that can be done is to invite the constituents to reform. This is how I attempted to carry out our politics. Our politics? Yes, including that of the profession, that of the Government, that of Parliament, and that of public opinion. One day, I said to General de Gaulle: "To be Minister of Agriculture of France, I must, first of all, give the farmers the feeling that I am their minister. It is only insofar as I accept this compromise in my innermost being that I can make them become aware of how, even though the French need them, their future nevertheless depends on the importance that France will agree to give them. In a country in the throes of urbanization, they must be convinced that it is not their electoral weight, and their demonstrations that will found the policies, but "common interest". He listened to me. He thought for a moment, and then went on to other things.

The Laws

So many new ideas and perspectives could not come to fruition without the dual political and legislative function of Parliament. These ideas and perspectives affected the forests, the landowners, the organization of producers, the capital for social development to develop agricultural structures, the schools. On the whole though, the debates opposed much less the Left and the Right, than they revealed a discrete complicity between a watchful Left and the-died-in-the-wool Gaullists, against unrepentant conservatives. This debate existed in the very midst of the government where Jean Foyer,[1] Valéry Giscard d'Estaign and Georges Pompidou represented three positions different from those of our traditional Right. To make progress, I had the determined support of Charles de Gaulle, and the voices of his "cronies"; more than many other mem-

---

[1](1921) Minister of Justice under the first four governments of Georges Pompidou (1962-67), Jean Foyer was a firm opponent of homosexual and gay rights.

bers of his entourage, they understood just how innovative he was, how antibourgeois, how resolutely attached to duration. I will give only one example of this. He had real knowledge of the forest, for having lived in it, and for often having gone for walks around la Boisserie; but he was unaware of the actual problems of managing it. He had to intervene twice when the status of the "public" forest and the "private" forest was redefined. Because they were subject to few constraints, private forests were, as the case may be, managed with long term-heritage concerns, or subject to accidents of speculation and inheritance. Conversely, public forests were not sufficiently required to earn a return on capital. Without being truly aware of it at the time, and encouraged by Michel Cointat, Merveilleux du Vignaux, Jean de Vayssière, I successively defended a two-pronged and converging transformation that tended to make the public forest more aware of the needs of the market and private forest, more respectful of the long-term, against some of my colleagues in the Government, against René Pleven[2] and Pierre Pflimlin in Parliament, against the young guard of private owners and forest developers. Two bills of law were necessary to bring this about. For me, this dual reform seems the best way to illustrate my preoccupations at the time: I had to begin not from an a priori, but from actual needs. In this case in point, the private sector needed some discipline, and the public one, some openings. We succeeded thanks to the President of the Republic, who imposed this two-fold movement on us. The effects of this political action continue to be extremely beneficial for the Forests of France!

One of my collaborators, present at the debate on the law creating "regional centers of private forests", and who was sitting very near a group of privatizing militants, heard one of them cry out in a loud and intelligible voice: "He should be killed!" He could only be referring to me. A little later on, having to choose a New Year greetings card, I decided to thumb my nose at the opposition, and reproduce the first page of a text adopted during the French Revolution that condemned to death anyone who would dare attack property. Not all of my correspondents appreciated the challenge, but by thinking about it and in a few years time, the vast majority of the landholders adhered to these policies that were dreamed up during a lunch between four people: Mr. Dubois and Vérilhes, President and Vice-President of the National

---

[2]French politician (1901-93), who was twice Premier of the Fourth Republic (1950-1951, 1951-52), and who also served as Minister of Finance, Economics, National Defense and Justice, during the Fourth and Fifth Republics. He is best known for his sponsorship of the Pleven Plan for a unified army. His efforts spurred the creation of the North Atlantic Treaty Organization (NATO).

Federation of Forest Owners, who had invited the Director General of
the National Forestry Commission and the Minister of Agriculture. Our
interlocutors were hesitant; Merveilleux du Vignaux and I pleaded our
case. In the heat of debate, I came up with the idea that the profession
should be organized like a "corporation", establishing rules, getting
them approved by the State and enforcing them through their own
association. Our interlocutors immediately agreed to this. I was to learn
later on, that the Vice-President was a Royalist: he did not fear, quite on
the contrary, the controls that were contemplated, he just rejected that
they should be applied by the administration.

Another debate on land ownership marked my ministry; that of the
"pre-emptive right" attributed to Societies for Land Development and
Rural Settlement. (SLDRS) They had the authority to grant land that
had been put on the market to small farmers, enabling them to grow
and to survive, and avoid the monopolization of farmland by large
landholders. The existing procedures had rarely ensured the regulation
of the land market: owners, buyers and notaries, in the main, got
around legal measures. Only the transparency of transactions and the
right of pre-emption could make these policies effective. The debate on
the corresponding articles was heated and intense. Had it not been for
the discipline of the majority of the Assembly, the articles would not
have been adopted. As for the Senate, it balked, amended, resisted, but
in vain. These regulatory policies did not yield the results we expected;
our plan of action was undoubtedly not what it could have been but I
am not ready, today, to think that regulation was not necessary. Because
of its patrimonial nature, land property remains, in effect, one of the elements that most crystallizes the structures and hierarchies of French
society.

The Law Governing Higher Education also had a chapter devoted
to the economic organization of producers. As all the farmers did not
want to be members of cooperatives, procedures and structures that
defined the regulations applicable to all, whether they happened to be
members or not, were put in place to deal with the quality of products,
and the conditions under which they were marketed. Hence, there was
no common administration, but common rules. Here too, the debate
was very intense; opposition was not ideological, it came from the
regions of large scale farming that did not want to see the rules that had
been established in 1936 by the Wheat Board reappear. I regret not having been a better teacher. Had it been better defended, the establishment
of associations of producers and the regulation of marketing would
have permitted fruit, vegetable, chicken, pork, cattle producers to avoid
the vagaries of the market that had become very speculative. It was not,
and could not be, a question of administrative control, but rather we
wanted to establish the mechanisms that ensured honest competition.

This ambition, which is in truth a modest one, is in fashion today. Such a waste of time!

The most original part of "my" Law Governing Higher Education is, undoubtedly, the Social Action Funds for the Development of Agricultural Facilities (SAFDAF) that I spoke about in the previous chapter. I would like to come back to its beginnings in order to stress its uniqueness in a country of "institutionalizing" law. Its fundamental idea is the following: every reform bears a cost for all concerned; they therefore oppose it. As this reform had a direct or indirect national impact, it was necessary for the citizens to be in part financially responsible for its implementation. The results obtained by the SAFDAF have been quite remarkable in spite of the "vagaries" it has undergone. In all of this, an idea remains that should become a rule: a reform that is useful for the national community, but difficult for those concerned, must give rise to a limited-term contract that shares the costs and the fruits of the necessary changes between the national community and those concerned.

From the standpoint of French agriculture that is used to functioning under the protection of laws passed under Jules Méline,[3] the influence and constraints of the market could disrupt everything, though in themselves they could not ensure the renewal that we desired so much. Hence, research and agricultural education needed to be encouraged constantly. To give an idea of the importance research had, Georges Pompidou, himself, came to the National Institute for Agronomical Research (NIAR) to liven up a work-session. The scientists told him about the progress made in the domain of knowledge, the problems raised by this progress and the means that needed to be put in place to take into account all the related domains in the field. A few weeks earlier, I took advantage of a major meeting of scientists and proposed a pact that was never recorded in writing, but that each one present had to promise to respect: "The State that supports this huge apparatus, does not expect you to devote all your time and efforts to it; on the contrary, the State understands that you pursue your own thinking with the tools it makes available to you. But the State expects you to devote eighty percent of your work to areas of research that it has identified as priorities. The Director General of the Institute has the difficult mission of ensuring the circulation of useful information and debates that arise from different viewpoints; he must also ensure this agreement is carried out and faithfully applied"… Long applause.

[3]Jules Méline (1838-1925) was Minister of Agriculture from 1883-85 and again from 1915-16. He founded the order of the *Mérite Agricole* and was one of the principal instigators of protectionist policies. He launched the idea of the *return to the land* which was taken up by the Vichy government during WWII.

But the reciprocal acceptance of society and change rests mainly on the system of education. The basis for a renewed agricultural education had been defined by a law Michel Debré and Henri Rochereau[4] had Parliament adopt. We had to apply it and make it work. Dozens of commissions were established that enabled us to modernize colleges and high schools, considerable effort was made to train teachers and to define programs. But when all is said and done, we had to explain the whys and wherefores of "agricultural education". Why did this specific "social" category, and only this one, need to have its own network or system that included secondary and higher education? Without a doubt, there were three reasons for this: the distrust traditional and even Legitimist[5] agricultural circles had for urban establishments; the families' refusal to send girls to school; the fact that agricultural activity corresponds as much to a state of mind as it does to a profession. I consulted widely and the answers I received stated more the obvious than they set out actual analyses. We had to get beyond this, or the dynamics would not be able to play out. The text passed by Parliament attempted to "secularize" the provisions that had been adopted previously. Defeated at the Assembly, I stated my opinions and my regrets and began to apply the letter as well as the spirit of the law. Everything hinged on the sum of the budgetary allocations I would be given to develop public institutions. They turned out to be generous, and agricultural colleges and high schools were soon fully occupied; a doubt was laid to rest: people knew that they did not happen to be good farmers just because they were born on a farm.

A few simple principles were retained: the creation of establishments that prepared adolescents for an agricultural and rural vocation, but did not lock them into this calling because there were large families. Yet, the number of farms was decreasing and those who would not stay had to be prepared for another way of life. We also wanted to accommodate young city dwellers who wanted to come to our high schools, since this mixing of cultures would be good for everyone. But agricultural education had to be opened up in a new way. Paul Harvois[6] proposed, had adopted and followed closely the creation in each school of a socio-cultural center where students could spend freely ten percent of their school time, and devote themselves to work and activities that in

---

[4]Minister of Agriculture (1959-61) in Michel Debré's government who was succeeded by Edgard Pisani.

[5]Partisans of the eldest branch Bourbon Kings, dethroned in 1830.

[6]Specialist of popular education who wanted to renew teaching methods and promote a socio-cultural form of education that would enable the young to take advantage of the evolution taking place around them. He also supported continuing education.

each school were defined by them, in consultation with the teachers. This worked extremely well. It was therefore with surprise and annoyance that I learned about the pressure that the professional organizations exercised for agricultural education to be only agricultural, that it train specialists, and only specialists who, in addition, were to be production oriented, and only that. I was annoyed because this pressure clearly showed the deep rooted ills afflicting one part of the agricultural world that thinks it can survive only by locking itself in a straitjacket. If one adopts that narrow attitude there is no justification for agricultural education. It can be justified fully, only if it sanctions, at one and the same time the specificity of a profession-state of mind and its opening up onto the world.

A final remark on this task of conceptualization and legislation: no provisions were adopted hastily, none escaped the rule we had established: consult, listen and attempt to convince leaders and militants who made their problems and wishes known. We never gave in to their demands. This was relatively easy, for once we had listened to them and informed them about what was possible and what was not; they accepted the decisions that, in the end, were adopted by the Council of Ministers, or the Parliament. Once adopted after consultation and debate, I do not recall a single measure that was contested by the various organizations in question. Some measures were criticized as being insufficient, and that is a good thing. I pity ministers whose flocks applaud without any reservations whatsoever; because they appease them by distributing subsidies, all they are doing is administering a placebo, or an analgesic. Hasn't this been the case in the past? But I also pity ministers who do not listen to their flocks, before telling them about and explaining the choices made by the government. Beforehand.

People, Space, Product. The Ministry

In France, there is no ministry like the Ministry of Agriculture. Indeed, none has the responsibility of "overseeing" such a set of different elements: products, society and territory, all together. In addition to the portfolios of administration and budget, to meet the challenge of the diverse tasks I was faced with, I felt it wise to set out three general directions that would both follow and drive public interventions related to "people", "products" and "space". This was done, while I remained in charge of the "The National Forestry Commission". I made certain I had a very restricted cabinet so that the members could not even hope to administer. The diversity of the tasks on hand led me to adopt two sets of practices: - weekly meetings over three hours long where I gathered around me members of the cabinet, and the directors general; each of us could intervene on any subject and, thus, take part in framing general policies so that we were aware of every aspect of them; - then, for one hour each evening, I had an open door policy, when I was available for anyone who had anything to say to me or to one of my colleagues. This made it possible for us to solve urgent problems quickly and to deal with annoying questions that always arose outside the usual secret forums, and what often happened, to meet with civil servants, experts who did not attend the long weekly meetings. Another rule: when my timetable permitted, I would go and spend two or three hours in one of the specialized departments of my General Administration, to go into detail, to explain decisions, to get to know the civil servants, to motivate them. One or two times a year, this desire to maintain contact led me to leave Paris, go visit the forests and meet foresters, their bosses, workers, owners and loggers. I had to leave Paris, since the forest is often forgotten in a ministry, where numerous crises arise around milk or fruit.

Because of this set-up I did not realize just how daunting my task actually was. When asked, I refused to accept becoming Secretary of State. I did not stop shuttling back and forth between Brussels and Paris, between my office, the Assemblies, governmental and inter-ministerial meetings, meetings with the leaders of the professions, between Paris and the provinces where I often went. To be able to do all of this, I traveled by helicopter. After I had done it the first time, and after having convinced the pilot to fly as low as possible, I came to the conclusion that no other means of transportation, no land travel could teach me as much as traveling by helicopter could. From that height, you can see all the details of the fields, the meadows, the forests and the houses. When I returned to the ministry, I often called on a co-worker to let him or her know what I had noted from the sky. One day, I made a bet with a local leader who was surprised that I had adopted such a slow means of transportation. I told him, that from the helicopter, I could evaluate the age of a couple living in a farm. He challenged me. We flew over his own district. He peppered me with questions. I was mistaken only one time out of five. The young women are the reason for this. The current generation treats its houses with greater care than previous generations did. A neat farm, a yard properly laid out, flowers, freshly painted shutters, are all signs of the "new generation". Deep in the heart of Brittany, where I carried out this experiment my reputation was widespread; though more as a sorcerer than as a minister.

Because of these trips, these contacts and talks, I became aware of the interrelation between space –including forests– products and rural society. It became evident that the petty wars that went on between agricultural departments, veterinary departments, agricultural engineering, national forestry commission, that were all part of my ministry, prevented us from addressing different, but interdependent needs, in a coherent fashion. The "local level", or "on the spot", is the only place where interdiscplinarity that is totally lacking in Paris, becomes an obvious necessity. I envisioned and studied the fusion between the trades and the reform of higher education, against almost everyone in my ministry. After lengthy debates, departments of agricultural Departments (DAD), as well as trades; Rural Engineers of the National Forestry Commission (RENFC), were created. A marked increase in the number of Polytechnical[7] graduates who chose "agricultural" careers was the first sign of success of this reform.

I will never forget these happy times without raising a number of questions I still have in mind: does the agricultural world in France not remain isolated from the rest of society by the simple fact that there is an entire ministry devoted to all its activities and all of its needs? Is it not dangerous to give the minister responsible for agricultural production the responsibility for the quality of the products; that is to say responsibility for food-supply? Is it not hazardous to give the ministry, and the profession, the responsibility for the environment, while the evolution of agricultural technology entails real risks? But at the same time, is it not totally absurd to think that problems of environment would be solved by defining and organizing protected zones? I still believe, today, it was a good thing forty years ago, when this huge transformation was taking place that all the problems of "French Agriculture" were dealt with, as part of a coherent whole.

Agriculture and Government

When I was at the Ministry, rue de Varenne, I became aware of the importance and the intensity of the relations that the agricultural world had with the State, the Government, the Administration, with the Legislature. This is not something new. It can be seen both in ancient history and in our own national history, from Sully[8] to Méline to Jean Monnet. This is also the case almost everywhere outside of France, in developing and more liberal countries: Japan and Switzerland, for example. The United States, zealots of free trade, continue to promulgate Farm Bills that are alternately electoral and strategic. We have to be aware of this custom, but at the same time wonder about it.

Political institutions cannot ignore agriculture, because agricultural policy is essentially a politics of the safety of the food-supply. Hunter-

---

[7]L'École Polytechnique, in Paris, is the most famous and prestigious school in France that trains engineers for the various departments of the country, along with officers for certain sectors of the army.

[8]French politician (1560-1641) who, as Minister of Finance under Henry IV, reestablished financial balance and gave pre-eminent importance to agriculture, in the French economy.

gatherers formed communities, and to ensure safety the first villages were organized around watering places or irrigation systems, the cause also of so many wars. From the very beginnings, to the present day, history teaches us that politics and agriculture are inseparable, throughout the world.

But means of intervention and practices have to be put into question, how, without damaging either, can the Administration intervene without upsetting the dynamics of the market? How, is it possible to ensure safety, that is to say, how is it possible to put in place the necessary regulatory systems without meddling with everything? Undoubtedly, by choosing to intervene only: 1) to take care of economic crises; 2) to face the consequences of serious climate changes; 3) to accompany the difficult mutations linked to common interests; 4) to keep alive entire, or essential, regions, where agricultural activity is weak for natural reasons; 5) and by that, maintain this immense, powerful and fragile "system" of agriculture in "working order", and make certain it can feed us. We have to take up this challenge, since it can lead to: 1) making the agriculture and food producing professions more responsible for what they do; 2) greater transparency in various interventions, as well as greater equity where they occur; 3) through public opinion, a better understanding of the diversity, as well as the importance of agricultural work, including a better understanding of the effort society must make with respect to agriculture, as its future depends on it. We have to rethink the collaboration between agriculture and State, and define more clearly the responsibilities of both, all the while ensuring that each has the freedom it needs to act. But there is always ambiguity in relations, where one of the two partners has all the money and power; the ambiguity is even greater when politics gets involved. I tried, but did I succeed?

Morality

I doubt if many politicians (but was I actually one?) experienced, or experience, all the joys I have. Was I even aware of this? I am not sure. In rereading a number of the innumerable speeches I made, I had the feeling that I was carried along by an historical situation that I had to take advantage of; and that, I could do so, only if I could surpass it. I had to propose to those who were at one and the same time prisoners of the situation, and beneficiaries of it, to project beyond, into the future, to pursue a certain vision of the future. There was a great deal of illusion in all of this, and we were the unwitting accomplices of a European strategy of commercial conquest. It is nonetheless true that if we had unmasked the stratagem and if we had refused to join the European Union, because other countries got a better share, we would have missed the boat. And, if we had found that Charles de Gaulle was taking too many risks in claiming the place in the sun he wanted for France, we would have missed the boat. If, instead of the unity of the agricultural world, we had preferred the quarrel of the ancients and the moderns, the difference that existed between those who had vast land-holding, and the modest farmers barely surviving, we would also have missed the boat. And, if we had forgotten to take into account the misery and hunger of the world, we would have betrayed our values. If I forgot today that, in this essential area, I had not accomplished one tenth of what I had promised myself to accomplish, I would be talking a lot of hot air.

Chapter IV

From Agriculture to Development

I was still in agriculture, seventy months after having arrived. I still found it very attractive. I felt "responsible" for things, and wanted to continue.

A few days after the elections of 1966, Georges Pompidou asked me to come and see him. He said I had accomplished a great deal, but the country needed me in another capacity. A very important ministry was going to be created, the Ministry of Public Works, where I would be responsible for public works, transportation, construction, urban affairs, and would have under me two Secretaries of State. I had to accept, for in entrusting me with this new position, the President of the Republic, and he himself, showed a great deal of confidence by a giving me such a wonderful post. What was I supposed to say? I asked for some time to think about this and, at the agreed upon hour, I telephoned to accept. In the meantime, I had learned that Edgar Faure had been chosen as my successor; in fact I was a pawn in a vast political game. I would learn later on that Georges Pompidou considered I had become unpopular in the rural world he needed to count on in order to become President of the Republic, and that he was already planning for this to happen.

I left without joy, but I soon learned that the sectors I was now responsible for, accounted for 62% of France's fixed capital budget. From housing to transportation, every French person "had to deal with me". This was a new ministry, and a new world. In fifteen months, I "established" the high speed train (TGV), by authorizing high speed on the railway lines, and succeeded in launching the Airbus! But in the morning, when I sat in front of my window, I thought about the fields, the meadows, the vineyards, the rain that nourishes, and the hail that destroys, along with the frost that burns. This is still true, today.

After spending a year in Public Works, I resigned in protest against the Government's use of "special powers" provided under Article 16 of the Constitution. This was a difficult moment for me, but then came the Student Revolts of May 1968. I thrilled at the call for change that came from the young. I climbed up to the tribunal of the National Assembly to speak about my fears, my reservations also, because I found the government's attitude timid. I provoked murmurs, when I dared say that a country is in a serious crisis when parents and children no longer understand each other, when parents no longer listen to adolescents. A few days later, I voted to censure the Government I had belonged to, and that I blamed for not having tackled or resolved the question of the universities and the problem of youths who were more anguished than impatient. This is undoubtedly the most dramatic day of my public life, but I simply did what I had to do. I became marginalized, belonging to no group, rejected by some, and inspiring some mistrust in others. I suffered through this. I soon became a member of the Socialist Party. Having no other qualifications except to serve the public I go through several years of uncertainty and difficulty, on a personal level. As a candidate for election to the Senate in the Haute Marne, I return again with joy to my department and the Luxemburg Palace that I had deserted, when I was appointed minister. I write a report on European agriculture, and the Presidential election soon follows. Almost as soon after being elected President of the Republic in 1981, François Mitterrand asks me to come and see him at his private residence in the Latin Quarter in Paris, rue de Bièvre. He informs me of my imminent appointment as European Commissioner of Development.

Once again, I came into contact with other agricultures, and with hunger. My mind wanders from the desert to the bush, to European debates on the Green Plan. I do not agree with the single options based on a single criterion, proposed by my colleagues at the Commission. I get down to work and on April 13, 1983, I dare circulate a document with the title: "Reflections on an Agricultural and Food-supply Policy and its Link with a Structural Politics and Regional Development". Already! But I learn about the incredible needs of the world, and the need for another kind of politics. This entails a responsibility that is more dramatic than any other.

I had not yet finished my mandate in Brussels when, at the beginning of December 1984, President Mitterrand gives me an urgent call. He must name a new delegate to New Caledonia. The situation is very serious, and he is looking for a person who has little to risk, and who is knowledgeable on problems in the area of public order and development. He asks me to play the role of peacemaker and reformer, in a territory heading for civil war. I leave immediately for this French territory. I find a difficult political situation, but also a dangerously unstable socio-economic state. I begin to work for peace, democracy and development. I learn a lot, I propose the notion of "independence-association", and make proposals of a political and developmental nature. I am named Minister, because no one accepts to defend before Parliament the law that institutes the evolution I propose. I am alone, and I am faced with a hostile Senate and a stormy minority in the National Assembly. I leave Nouméa, the capital of New Caledonia after an exhausting year. During this time, I was able to reestablish peace, I negotiated a new status for the territory that was passed by a referendum, and I gave the Territory a system devoted to development and to the organization of elections that unfold in a very peaceful way. I hand in my resignation to the French Government since, for me, it is not a good thing for a Territory to have its own minister. François Mitterrand appoints me technical advisor to the President, and I take on the tasks that I choose to. As I will soon be seventy years old, I organize my "retirement", I publish a few articles, and I am soon asked to preside over the International Center for Advanced Research on Mediterranean Agriculture, then to become President of the Institute of the Arab World. For a few weeks I presided over the preparatory meeting for the General Assembly of the United Nations, devoted to Africa. I would like to touch on what I retain from these three experiences (in Africa, New Caledonia and around the Mediterranean). They taught me about other forms of agriculture. They represented an immense problem which would increase uncertainties in the world if it were not soon solved.

Sub-Saharan Africa today has five hundred million inhabitants, and will undoubtedly have a population of over eight hundred million before 2050. Added to the rhythm of migrations to the cities that I noted before, this increase in population creates a problem so great, that we cannot even imagine a solution to it. It truly represents a danger, and there are only two ways to deal with it: control demographic expansion and encourage agricultural development, because it slows rural drift from the land and brings about global development. No matter what the immediate, or future results, of a policy to control demography, agricultural development will still have to deal with the problem on two fronts: to enable countries to decrease their dependence for food, and to use their resources not to import rice or milk, but instead, machines and tools; to slow urbanization that is exploding. I will return to these issues later on.

The problems that arise on the Mediterranean perimeter are of a different nature. Two essential features characterize Mediterranean agriculture: the lack of land and the lack of water. Except for a few plains, irrigated by rivers, the entire basin is hilly, rocky or sandy, and gets very little rainfall. Vineyards thrive here, though, the wine is better than the one produced in silt soils; olive trees also thrive here, sheep and donkeys are agile climbers. But taken all together, all of this does not feed the numerous populations that live, and are increasing in this region. The food deficit is increasing everywhere, and cannot be met by irrigation. There is not enough water, and this state of affairs will continue, especially since human and industrial needs are multiplying with the evolution of customs, the development of tourism, and the multiplication of food processing plants. The battle for water is crucial. At the same time, in the Mediterranean basin the soil is being damaged by erosion, the encroaching desert, over-irrigation, or the abusive use of fertilizers, that is threatening; and everything must be done to stop this.

And urbanization is gnawing away at, or devouring, the best lands. The politics of the land that must be protected, and of the water that must be saved, is the politics of the future. This was true, even fifteen years ago, but it is all the more so today, because changes in climate, and ever-increasing droughts, have caused great damage over the last years which imperils the economical, political, and social balance of certain countries. These two areas of concern are all the more serious, because the scarcity of space and the scarcity of water make self-sufficient food production more difficult. Except for two or three countries, the Mediterranean basin cannot ensure the subsistence of its people. They have to buy necessary foodstuffs, and concentrate on the production of oil, wine, and fruit, all the while developing light industry. One begins to dream about a "pact" that would make the Mediterranean the vegetable garden, the orchard, the vineyard, the oil cruet of Northern Europe that needs to brighten up its table, because it has an excess of cereal, milk, sugar, and meat that it does not know what to do with. It is absolutely necessary to "think" about the conditions and means to bring about this pact that would permit both shores of the sea, both adjacent continental spaces, to attain a degree of fulfillment. Land and water; the garden and the granary; an agriculture and an agro-food producing industry that creates jobs; a pact that guarantees the South equitable relations with the North; here are four themes one must think about, from a political as well as an economical and social perspective. These problems concern all of us more than we would like to believe. Migrations that are occurring force us to concentrate on them.

From my stay in New Caledonia that was so rich and eventful, I shall retain only two anecdotes, because of the "philosophical" lessons they contain. One day, when visiting a commune on the central plateau; and walking alongside a plot of land that was very well tended and very productive, I came across a plot of fallow land. Thinking it had been left fallow, I inquired about farming practices here. As a matter of fact, this was not the case, and the farmer answered that he did not have to cultivate more land since the land under cultivation fed his family; it provided surplus produce that was sold and enabled them to buy groceries, clothes, and drink beer on Sunday! During this same trip, I saw a nice parcel of land planted with trees. I asked about the owner. "Here, I was told, the land does not belong to us; we belong to it."

Self-sufficiency and productivity, usufruct and property... where does wisdom lie, in all of this?

Chapter V

Articles. The Seillac and Bruges Groups

I had a remarkable experience during the years that I have been reminiscing about. Pierre Mendès France[1] asks to see me. We were well acquainted for we saw each other often during the Student Revolts in the spring of 1968. Without any preamble whatsoever, and because of his extreme fatigue, he asks me to replace him as member of the Commission for the New International Order, presided over by Willy Brandt.[2] Some of the members of the Commission were quite outstanding, and it met six or seven times around the world, trying to assess the situation and trace future avenues to be explored. I do not miss a single session, and try to make the voice of agriculture heard. The group is motivated by a free trade ideology, and its foremost preoccupation is the opening of international commerce. I remember the "arguments" I had with representatives of the Third World who were desperate about the actual state of affairs and expected that liberalization would solve all their problems, and that they would be able to invade European and American markets. In spite of Willy Brandt, who is a major figure, liberal ideology carries the day; and agriculture is downgraded and treated like all the rest: there is hardly any difference between trading food for subsistence; and textiles, boots or equipment. During these meetings, I am able to meet personally with Helmut Schmidt, who was the Social-Democratic Chancellor. He tells me that he gets along very well with Valéry Giscard d'Estaign, but through experience he ends up thinking that this liberal is more for centralized state control, than he himself is. He tells me it is a question of culture; "you are Jacobins, and

---

[1]French politician (1907-82), who was a militant during the Resistance and President of the Council of State, holding the portfolio of Minister of Foreign Affaires. His government negotiated an end to the war in Indochina (1954), and had to deal with the Algerian War. He negotiated the treaty for the independence of Tunisia. First Vice-President of the Socialist Radical Party, he was one of the initiators of the Republican Front (the union of the combined non Communist left). Member of the unified Socialist Party (1959), he wrote numerous books and was considered an intellectual guide of the French left.

[2]German politician (1913-) who was a militant during his youth of an extreme faction of the Socio-Democratic Party. From 1950 to 1957, he was deputy for Berlin in the Bundestag, then mayor of of West-Berlin from 1957 to 1966. He was elected Chancellor in 1969, and continued his policy of rapprochement between the West and Communist countries. He subsequently won the Nobel Peace Prize in 1971.

believe in the State. We don't!" I tell him that I am not certain that as a German liberal, he is truly not interested in agriculture. He places his hand on my arm and confesses all the while not pleading guilty; he gives me reasons that are not all political ones. They just happen to be Politic.

I am no longer in International Affairs, but for many people, I remain the former Minister of Agriculture. In 1972-73 I dare openly state that European Common Agricultural Policy must be changed because, since it has attained its objectives, it must no longer encourage only developing production. In 1976, I was asked by Pierre Joxe[3] to write a paper for use by the Agricultural Commission of the Socialist Party. I will quote a few brief passages: "We must promote a world-wide policy of food production and distribution, and spearhead a res-olute agricultural policy...The CAP is driven by a policy that favours productivity, in spite of the fact there is no such thing as a policy that is not at one and the same time global and economical, as well as region-al, social, ecological...If it is true historically that the construction of Europe could only begin by the organizing primary markets, at that time governments made a serious mistake by not giving themselves the means to deal with agriculture in all its complexities...If Europe can't have a global agricultural policy, if Europe doesn't become a substantial geopolitical reality, must France continue to be part of it? We have to ask that question for, from the point of view of agriculture more than from many other points of view, stagnation, the break up of Europe, raise daunting problems... The balance of risks and opportunities that Europe represents must come about, it must be made public...The Brussels' experience teaches us that only an unshakeable political will, supported by rigorous analysis and a coherent project, can triumph over contradictory interests and ideas." I was ready to develop my analyses and my projections further, but I was tired of this and it was like talking in the wilderness to put into question a policy that seemed to be obvious and which, after having been reformulated in terms of ever increasing productivity, was ardently defended by professional organizations. It was no use reflecting on the agricultural problem, because it had become the prerogative of conniving bureaucrats: pro-fessionals, ministerial, national, from Brussels or international. I chose to step aside.

---

[3](1934—) Son of Louis Joxe, minister under Michel Debré and George Pompidou, who was considered as one of the left-wing supporters of General de Gaulle. Pierre Joxe was in the cabinet of the Minister of Foreign Affairs from 1967 to 1970, and became Deputy Secretary General of the Convention of Republican Institutions in 1970, then National Secretary of the Socialist Party in 1971.

After a long absence, I intervened on several occasions, twenty years later. In September 1994, before the general assembly of the "International Association of Food Products and Agro-Industrial Economy" then, in April 1995, in an article published in the newspaper, *le Monde Diplomatic*, with the title "For the World to Feed the World". I defended what was obvious: If one takes into account the increase in the population of the world, and the depletion of natural resources; it is not obvious that all the people of the world can be fed; the Western World must counteract its own wastefulness; all of humanity must struggle against its propensity to destroy the environment. We must put into question the demographic explosion that makes all of these problems so difficult to solve. We must use our excessive agricultural production to participate in the struggle against hunger, but at the same time to encourage development.

Because of these public stances, I am asked in the autumn of 1995 to chair the "Symposium on the 50th Anniversary of the World Congress on Food", that took place in Quebec City. While summing up several days of very rich exchanges and debates, I state in my own name that: "We have experienced a bipolar equilibrium where the problem of safety was the most important of all. This division into two parts, is now obsolete, and we must interrogate ourselves about the future balances of the world; not the balances of armaments but those of demography, of wealth, those of access to means of survival, of life, and those of peo ple realizing their potential…(The world) has entered into a period of super-fusion, where everything is uncertain, where everything is almost risky. Now all of this will crystallize once again over a period of time that can be quite long. It is our duty to intervene, so that crystallization occurs around a certain number of values, objectives, necessities. If this doesn't happen, then the crystallized world will necessarily break up, because it has not assumed responsibility for its own problems. The system of the world will explode, because it has willingly ignored certain urgent demands…" In December 1966, with Bertrand Hervieu, we asked the question in *le Monde*: "What agriculture should Europe have?" We affirmed that: "in giving preference to networks over territory, European agriculture is part of a movement to create a break between the economy, on the one hand, and territory and society, on the other; this is a break in the traditional balance that is at the origin of a new way of life, and civilization".

These numerous interventions provoked a number of inevitable responses. I was asked to undertake a global and methodical reflection on the problem of agriculture, situated within the perspective of the last article cited. With Bertrand Hervieu, we brought together, first at Seillac then in other places, twenty or so people, (farmers and non-farmers) so that we could set up a think tank that thought in contradictory ways. I published the contents and the conclusions of these meetings in a book that appeared under my sole responsibility. It analyses the thinking of this group, where each of the participants could identify his or her ideas. I gave a résumé of the Conclusion in the Foreword: "French and European agriculture must be productive and competitive in an international market where they ensure our interests; also, in their own internal markets, since they guarantee our self-sufficiency and our autonomy. But they must treat our territories with prudence, since the latter make up, at one and the same time, a significant portion of our civilizations and our societies, as well as contribute to our lasting equilibriums. French and European agriculture aspires to this, and must do so by means of an implicit pact and a well-defined policy. France and Europe must state clearly, once again, what society, what territory they need, and what means they accept to devote to this planning by objectives".

How could an independent group, better than the professional organizations or the responsible politicians, have sketched the directions for research in this area, and defined the elements of a solution? Tackling these issues leads to a radical questioning of the institutional system that imprisons us, because of hierarchical traditions, private preserves, and specializations that ossify it. It simply transforms debate into an inter-play of roles. Our intervention is contested, because it does not come from an accredited organization. The little book I put together gives an answer to this: "The autonomous moves, that do not have any of the strategies and ambitions of the Seillac Group, rest on several very clear certainties:

—No matter how militant one happens to be, a human being is still a human being, free to think and speak as she or he wants, as long as they do not involve the organization they belong to. When we disregard this rule we paralyse fruitful exchange, and this leads to so many conventional and useless speeches. This is what legitimizes "stereotyped formal language".

—In a democracy, the problems don't necessarily belong to institutions, or to organizations; they belong to the immense domain of free discussion...

—The agricultural problem, the crises of agriculture and rural environment, are part of a general crisis affecting modern societies; and if we attempt to tackle them in isolation, we basically refuse to deal with them. It is in their common interest for the agro-rural world and for society as a whole to confront their situations, and their difficulties.

—There exist a huge number of men and women who, by preference and through a sense of responsibility, want to express themselves in a free and disinterested way on the problems facing them. They also want to contribute to awakening those who are concerned by these problems; as well as to contribute to the solution they hope for.

—Do people not see that democracy would be more efficient, and society healthier, if informed opinion framed and inspired the workings of institutions and the media through questions and enlightened views? A living democracy imprisoned in the halls of deliberating assemblies."

These few sentences are not pure and unadulterated literature. Those who met with the group, rarely advanced the positions of the groups they belonged to, and if, in each intervention, we could feel traces of belonging and adherence, we could especially feel the desire to inform the debate with an analysis that invited other analyses and other projects. Under the "cultural" influence of city dwellers, of macro-economists and scientists, during sessions, or at meals taken together, during walks in the forest, in front of a burning fire in the fireplace, tongues were loosened, and old quarrels gave way to authentic attempts to pursue solutions. Those that occurred between farmers and non-agriculturalists, facilitated the reconciliation between opposing agricultural visions. These are the agricultural and general conclusions of the Seillac Group that after it became European decided to meet in Bruges, where the directions that had been traced continue today. It has just published a manifesto whose title gives us hope: "Agriculture: A Necessary Turn".

Conclusions: Anxieties and New Hopes

Because of its vocation to feed, agriculture has always and everywhere been the object of special attention by politicians. It is the obsessive and desperate preoccupation of all Southern countries. It is also that of the Northern governments, since none of them has ever attained a state of stable balance between production and consumption, nor even a stability of world prices that protects it from the "peasants' anger". The natural irregularities encountered in producing agricultural goods necessary for the survival of the human species are the reason for the social crises that affect producers who are devastated by surpluses; or consumers where increases in prices have an impact on their standard of living. Forced to face such crises, all governments intervene, except in those countries, and there do exist some alas, where a person who "dies of starvation" is considered simply to be a victim of fate. Governments also intervene in countries the size of continents, with populations of over one billion people. Peasants represent the major portion of these populations and, to feed them, half would have to migrate to cities that are already overcrowded. What can we do?

People are undoubtedly thinking about creating a place of their own for agriculture, and rural space in our vital balances. But the objective of production was favored over all others. In the meantime, people began to worry about the environment; and this was justified because of agricultural pollution; and they worried about the food-supply, which is the inevitable consequence of forced production. Such fear inevitably occurs when performance is favored over all else.

I was also a productivist...in the past.

This was in answer to the situation at the time, and I don't regret it. I have an obsessive need to see further progress made here: we need to integrate all the variables, and assume the new complexities: we need to feed human beings safely, to take part in modern civilization, to contribute to national economic balances, to offer our rural societies a real future, to protect nature and to enliven rural space, to end the hunger that affects hundreds of millions of people, to assume, to pursue progress by filtering it through new forms of behavior. We must from now on: 1) find an answer to the problems of our times; 2) be ready to find other answers by forbidding ourselves to consider each answer as being definitive; and 3) be constantly on the look out for an "eternal" order that saves us from the temptation of shaping an era of post-nature for a post-human being.

We must question everything, so that confidence will be rekindled between a demanding urban society and farmers who, as free and responsible beings, ask only to earn their living from their work and their investments.

I am beginning to think that we must organize our politics, (French, European, World), starting with the explicit recognition and respect for the two "models" that take into account the immense diversity of environmental and social conditions, of needs, beliefs and cultures. How can one characterize these two symmetrical "models" or, rather, these two references? I would like to state the following: the first model, or reference, favors market production, all the while respecting nature and rural society; the second, accompanies the often countless masses of the rural populations that are looking for work, the environment, and production that needs to be ongoing. These two "models" are necessary and complementary and illustrate the infinite variety of balances that exist between the economic function and the socio-spatial function that stem from each of them. No research and no politics can favor one to the detriment of the other; for it is in, and from this diversity that the world can live better, it is by respecting this diversity that the world can achieve its unity.

And now I question the very meaning of the times we are living. Do they foreshadow an inevitable post-modern future; do they not instead lead to new Middle Ages that have not yet found the ways leading to a Renaissance? If Modernity wanted to impose all of its discoveries, its model of consumption and the reductive market vision that motivates it; if, enamored of itself it wanted to mould people and treat nature badly, I could not prevent myself from considering it nothing more than a barbarous adventure. We can do better than this.

Part II

Assessments, Prospects, Choices:

Defining a Policy

Introduction

The last half-century has been marked both by historically unprecedented demographic increases and increases in revenues. Three major dynamics have enabled agriculture to deal with these: the market and the farm-produce industry; technology and science; and the agricultural world itself. The European Economic Community founded its power when it adopted them; if it had not taken an original stance with respect to these dynamics, it would have lost its relevance and even its very raison d'être.

In the first part of this book, I retraced the experience of a man who knew nothing, or almost nothing about agriculture. In 1961, I was given the task of both spearheading French agricultural policy in Paris, and negotiating and soon applying Common Agricultural Policy adopted in Brussels on January 14, 1962. I navigated from one theatre of operations to another, constantly and straightforward, by reorienting a policy, by drawing my arguments from another: French agriculture was by far the most important, and the most diverse of the six nations in question; the European market provided the outlets the country greatly needed, but had cared little about until then. This fortunate ambivalence happily determined its action. But it made the market a proponent of production, productivist. There was no time to analyze and evaluate, to recapitulate and open new perspectives. There wasn't time to say why Europe and the world can neither be content with these principles, nor continue their practices. Later on, I will propose another policy. Hence, Part II of this book will have three chapters: the first will explore the "dynamics" of "productivism"; and the second, its "results"; the third will describe "the search for a policy".

No matter how objective Part II attempts to be, it certainly is not neutral. It begins with a simple idea; the necessity of satisfying needs after having identified them. This approach is less mundane than we could believe, for it is almost impossible to speak about agriculture in institutions and in professional organizations without analyzing the existing texts, the compromises that have been worked out, the promises made; it is impossible to speak about agriculture, impossible to break with the hard and fast alternative between the two terms: "We have to" and "it's impossible"; impossible to escape from an obsession with the market, and therefore to respect the trilogy that should structure all agricultural policy around product, people and space. Analysts, and negotiators, are prisoners of their own constructs; in their eyes every new thing that is true, is incongruous. When facts force experts to adopt a specific train of thought; they indulge in a few contortions and pro-

pose makeshift solutions that are laughable when they are not catastrophic. The idea of reform is foreign to them and they have given it up
a long time ago. Between "we must" and "we cannot"; desultory hope-
lessly trite reforms are adopted that convince no one. This is how we
put up with Common Agricultural Policy: its object is to last, but without understanding why. It has reached this state because it has not been
rethought after having been successful. It should have been rethought,
because it succeeded and actually changed the reality it was dealing
with. To enable it to regain its former relevance, all we have to do is to
define it so that it encourages agriculture to respond to today's needs.
Indeed, one doesn't undertake reform to make things last, but to make
them work. One doesn't reform a policy with the idea of finding a compromise between new needs and acquired rights, but with the will to
establish a new balance between the needs of global society and the productive activity and the living reality of the agricultural world, even if
it means changing it later on as the need arises.

The CAP has become unpopular, even among those whose mission
it is to help. They hang on to it like a lifebuoy, because it hands out subsidies. It is unpopular with city dwellers who, when confronted with
unemployment, do not understand why farmers receive the help they
never did themselves. It is also unpopular with farmers from countries
that will become future members of the Union, with the United States,
with the Third World. It is the preferred target of international institutions. It satisfies no one. It has lost all legitimacy in the eyes of economists since Europe can supply itself on the world market. Those who are
empowered to administer it are preparing on the sly to revise it again,
which will lead to its demise. It was created after the Second World War,
when people actually remembered rationing; it focused on production.
It is not adapted for a period in history, when safety, the quality of food
and the environment preoccupy public opinion, more than subsistence.
Nothing can restore it. It needs to be reinvented from the ground up. But
is this possible when the European Union is taking the shape of an association of countries with limited responsibility; more concerned with
their diverging interests than their common destiny?

This is necessary, for the stakes of a new policy do not only affect
the future of agriculture, food production, rural life and the environment, but also affect the future of the Union at the very moment it is taking in new members; when common currency has not been unanimously accepted; when it clearly shows its weaknesses and its divisions
during the most serious ordeal that the world has gone through for a
long time; and, at a time when the world is in disarray. Through its existential function, agriculture provides the European Union with a chance
to show that, above and beyond serious difficulties, Europeans have the
will and the ability to act and live together. If we achieve this, then the

rest will come. If we fail, then Europe's destiny will be in jeopardy because if it cannot resolve the essential problem of security and balance; it will have failed to define its own conception of modernity. If Europe seeks modernity through diplomatic agreement and is unwilling to propose an original vision of the world, then it runs the risk of soon becoming a small-time player on the world scene.

Chapter 1

Dynamics, Results, Assessments

From Rationing to a Common Agricultural Policy

On the eve of the First World War, France is self-sufficient and the primacy of agriculture is not contested. Just after the Second World War, in 1945-46, France imports 26 million bushels of grain, and 300,000 tons of meat. Because it is necessary to restore the country's balance of payments, agriculture has to become an exporter of food. Agriculture is therefore given a place of choice among the founding objectives of the Plan for Modernization and Development. As early as 1950, agricultural production regains its pre-World War I level. The status of cooperatives, the bank of the Crédit Agricole is reformed, the National Institute for Agronomical Research is created, and agricultural education is also reformed and developed. While progress continues, exports are developed, and prices plummet. Decrees are passed to organize the markets. Farmers demand the "parity" prescribed by the Law Governing Higher Education. The crisis is not due to fluctuations arising out of a few economic conditions, but rather stems from the logic of agricultural development that is considered necessary. It explains the role that France demands for agriculture in the European Economic Community that is being created. It explains all the efforts made, and one discovers that it would have been much easier to create an abundance of food, than to manage it. No one notices the inherent contradiction between economic objectives based on increased production, and the desire to enable the largest number of families to survive on the farm.

But driven by political and inter-professional will, the fifteen years that separate the Armistice from the adoption of Common Agricultural Policies, enabled French agriculture to overcome its technological inferiority; to convince professional organizations to abandon their corporatist attitudes; to ensure promising prospects for the farm-produce industry and commerce; to make certain that foreign countries brought produce from us, other than wine and cheese. During these fifteen years, French agriculture became a significant economic reality. Two statistics bear this out: the production of cereals went from sixty one million one hundred thousand bushels in 1945, to over one hundred million fifteen years later. The same holds true for all other aspects of food production. The Common Agricultural Policy was the perfect answer to an unwieldy prosperity.

The CAP came about through exhausting and fascinating negotiations, because each of the participating countries was torn between
defending its immediate interests and opening up new agricultural,
economical and political perspectives as well. Three principles were at
the origin of this new reality: the free circulation of products within the
Community; the protection of borders; financial solidarity. Because
French agriculture ran a surplus, it occupied a position of choice with
respect to German and Italian agriculture that, on the whole, run a
deficit. But things were arranged so that the "integrated" agricultures of
the six countries benefited from the policy in question. They would
have benefited much more so, though, if they had taken into account
the evolution that was occurring in the area of cattle feed, and if the production of feed had been organized in the way cereal production was.
Nevertheless, European agriculture prospered and became the second
exporting geographic entity in the world. To such a degree that it raised
problems for the farmers of the Mid-West, the Canadian, Argentine,
New Zealand Plains, and would become one of the most contentious
issues in international negotiations.

Without a doubt, agricultural policy has created an internal and
external dynamics that has turned out to be beneficial to the six
European countries, enabling them to develop in a way they could not
have within their own national borders. Yet, this policy has given rise to
two problems which, without putting into question its merits, nonetheless changed its significance: it is unevenly protective; unevenly "generous" with respect to products, farms and regions. It is undoubtedly
more beneficial for the European Community, than for its farmers; the
policy gives the Community a political sense that it would not have
without it; it has enabled the Community to position itself in international economic trade in a way it could not have done without it. But
what's the use of complaining when it is obvious that without the CAP
European agriculture would have found it much more difficult to modernize, and to benefit from a continuous dynamics?

The Market and Firms

After the Second World War, the Common Market Policy served both as
an incitement and as a guarantee for the essentials of life that were, in
effect, positive constraints. Without a doubt, the three dynamics that
spurred on development were the market and the firms; the technologies and sciences creating seminal innovations; and the "farmers" and
"peasants", who were the actors on the land. I would like to examine
these dynamics in turn without, nonetheless, claiming to say everything
that should be said about them.

When dealing with the market and firms, I will refer mainly to

French facts and figures. In short, thanks to an industrial and commercial complex chaffing at the bit to modernize, the farm-produce industry was able to make up in part for our national deficit. Here are some facts: Between 1959 and 1975, the farm-produce industry went from 14 to 16% of our exports, and decreased from 29 to 13% in terms of our imports; the cash margin to cover our goods, progressed from 51 to 114%. This happened in spite of an increase in our population from 45 to 53 million inhabitants, and a no less significant increase in individual food consumption; in spite also of importing specific tropical goods; in spite finally of the place the import of cattle feed holds in our assessment. In examining these results, our "exports" to our partners in the European Community play a very important role. What would we have done without them? But all of this is now being questioned. As a result of a particular vision of the globalization of the world, we are invited to open our borders not to agricultures that happen to be more modern or more efficient than ours, but to agricultures that can produce food more cheaply than ours, because of their natural conditions or their social practices. Our results are often better than those of the Americas or Australia, but their conditions of production, land or labor, cost less than ours. What can the world-wide future of our agriculture be, when market concerns tend to be more important than diverse agricultural realities; when concerns for the future no longer inspire the respect for all that can be produced in a world, where needs are increasing faster than the objective capacities of production? This future is dependent upon the food-producing industry and commerce; they are the motors of development of a sector which has enabled France to become the second exporter in the world, and has become the first French manufacturing sector, even ahead of the automobile (130 billion Euros versus 86). We can sum up the importance of this sector as follows: of the 425,000 products offered to consumers, 100,000 have existed for less than a year. At the same time, this sector has developed its technological, commercial, exporting and financial capacities.

Thus, necessity, the CAP, the modernization of agriculture, and the appearance of surpluses associated with the evolution of the population, of revenues and of consumptions, brought about the development of a powerful industrial and commercial "system of food-production". We went from self-consumption, and short circuits of agricultural foodstuffs, to a system of production of mass products, transformed, standardized which circulate through long chains of intermediaries, and that are renewed each day to meet the tastes of the clientele. In terms of finance, employment, value added, industrial production and sales have become more important than agriculture itself. Within the framework of massive contracts, the former would willingly have a tendency to relegate agriculture to a secondary position. Large cooperatives do not oppose this necessary but formidable tendency; for it is obvious that without the huge firms with world-wide networks, our agriculture would have never been able to attain the levels of efficiency and production it has.

All analyses confirm this. In developed countries, the share of the price of agricultural products in relation to the final price of the food products continues to decrease: on the average, when the consumer buys food for 100 Euros, 20 to 30 go back to producers. As far as beef is concerned, of the 100 Euros paid by the consumer, the producer receives 35, the "intermediaries" 21, and distribution 44. Over the last fifty years, the share of food in French families' budgets has diminished from 31 to 17%. Food costs have been increasing much more slowly than revenues. But the qualitative and health expectations of consumers have been increasing, and a split has been created between "rich" consumers and the others. We are evolving towards a segmentation of the markets, with differentiations in tastes and degrees of preparation. The food-production system has developed to such an extent that the commerce of unprocessed products has a tendency to diminish and become marginal in international markets. But 90% of the food-production industry world-wide is controlled by five huge firms.

As the indisputable motor of development, that is both surprising and beneficial, the non-agricultural part of the food-production sector has a dominant position. Firms are developing hold-over "farms" that they shape at will to the detriment of the environment and the rural fiber that depend on them more and more, as years go by. The suppliers of agriculture also play a considerable role; whether in machinery, seeds, fertilizer or pesticides, but by encouraging consumption, they run the risk of derailing modernization and overwhelming small units with useless and even dangerous purchases. Nonetheless, their intervention has had a positive and decisive effect.

Technologies and Sciences

At the end of WWII, two things defined French cattle: their low yield of milk, and their extraordinary heterogeneity. On the average, cows yielded less than one thousand liters of milk per year, and herds were a mixed lot. The production of milk had to be increased by favoring breeds that had proved themselves. But it was impossible to slaughter tens of millions of crossbreeds that filled our pastures and stables. We chose to proceed, step by step, and to improve the stock by selecting bulls. But there were not enough bulls; and cows or heifers had to travel too far to be inseminated by a selected bull. This was not a problem: by selecting bulls, by collecting sperm, and by dividing it into fertile lots, by freezing it and inseminating cows in stables, centers of artificial insemination brought about the impossible: instead of inseminating a few dozen cows per year at great cost, a bull could inseminate a few thousand! The progress that occurred with regard to corn is equally impressive: at the end of the war, corn was cultivated only in the valleys of the Adour and the Garonne in the South West of France; twenty years later, hybrids appeared in the valley of the Loire; and twenty years after this, corn thrived in the valley of the Rhine and everywhere else in France, with surprising yields.

At the beginning, the National Institute for Agronomical Research played an extremely important role in this. Its research in the selection of plants led to the creation of new varieties that were so successful that, in turn, it led to the creation of an extremely dense network of small and medium sized businesses and industries specific to France (the second producer and third exporter of seed in the world). The research it did carried over to many other domains: genetics, plant physiology, agronomy. This research enabled widespread circulation of scientific culture and technologies within the professional community; its technical and cooperative organizations; as well as among farmers, themselves. To such a degree, that although farmers were hesitant at the beginning, they soon became eager clients. An actual industrial stampede occurred: embarking on biotechnology, firms patented widely; from a few new "patents" per year during the 1970s, to a few dozen in the 1980s and to 100, 200, then to 300 new varieties a year! The fact of the matter is that competition is more dynamic than "progress of knowledge", and even more than publicly funded teams of researchers can be. This raises a problem that should be considered, since the concern for profit even if it happens to have a stimulating effect can have negative consequences. I will return to this later on. It is nonetheless true, that whether research happens to be publicly or privately funded, it covers all domains; that of pathology and therapy; the prevention of plant or animal diseases; the working of the land; mechanization, from tractor, to harvester, to sower, to milker, is drastically changing the farmer's work, and massively increases the return on human labor. In short order, the value of the units of production is more knowledge and technically based, than they are on the land! This is the case with the food-production industry which in a half century went from the conservation of agricultural products, to their transformation and, quickly, to the frenzied renewal of products that are "ready to consume". Scientific and technological development have enabled, and still promise to lead to, a further reduction of the house keeper's and the farmer's work load, but not only has it dramatically disrupted the life of the former, and the work of the latter, it has also disrupted rural life and created problems for the environment.

"Landholders" and "Peasants", Farmers

Agriculture is at one and the same time a sector of production and a social universe. For the great majority of today's remaining 680,000 farms (there were three times as many in 1961), agriculture remains an individual project. In each farm, the person responsible works like a laborer and oversees his/her budget and plans; much like a company manager. In spite of having a production value of 420 billion francs and 61 billion francs in commercial surplus, it still remains a family affair! The overlapping of family/profession remains strong. One third of the operations are run by women, and this hardly changes anything. Male celibacy is still prevalent, and the cohabitation of generations persists. Large families are much more common than in the city, but there are fewer and fewer of them; marriage partners are chosen in the "profession". The social base shrinks from generation to generation, and 80% of farmers' children take up a non-agricultural profession. The lifestyles of farmers' change, but through necessity as much as through choice, it remains very different from the city dwellers'. Fifty years ago, houses seemed to come out of the Middle-Ages, but today they are equipped with computers, refrigerators and television sets. Most houses have running water and paved driveways. Little by little, and at the risk of becoming a house in the suburbs, the family home is located further and further from the farm. The telephone is used and abused, and farmers go to town for any old reason. There are hardly any agricultural shows anymore, and farm fairs are becoming less and less common. People read a lot, and they keep their own accounts. There are not many farms where more than one person works fulltime. In order to make ends meet, the wife or the husband work elsewhere. On large farms, the owner and his family often do not live on their land; they have an apartment or a house in town; the owner comes when the seasons and work demand it. Farm work has changed.

But let's move on, for agriculture is infinitely diverse. I would like to examine four models that correspond to economical realities, to different ways of working, to cultures and ways of life that, although they have come closer together, nevertheless remain dissimilar. The first model consists of plant-raising farms that focus on "raw materials", and provide silos, or cereal factories, with oleaginous grains, or sugar beets. They can be compared to mines, but whose resources must be renewed on a yearly basis. Their problems are proportional to the extraordinary progress they have made. In spite of their performance, they can survive only by means of subsidies, because of world market prices. The second model corresponds to meat raising industries, whether they happen to be pork, poultry or cattle. Allowing for exceptions, the seasons hardly matter for they are governed by time constraints: they have to deliver chickens that are seven weeks old, pigs that are twenty, and calves to be butchered in thirteen or sixteen months' time. They force-feed their animals: during their last three or four months of "life", calves put on almost a kilo per day. Calves, pigs, chickens: the products are standardized. This is what can be called industrial life! The third model is made up of milk farms. Milk is delivered like a raw material, but there is no need to rush, since the cows stay on the farm for years on end. Everything is mechanized, but each cow gets "personal" attention. The fourth model is a grab bag, but it is a bag of immense richness; it contains vineyards, where "appellation contrôlée" are at the top; orchards and vegetable gardens, the Beaufort and Laguiole, where "appellation contrôlée" cheeses are made; there are the thousands and thousands of little farms that raise ewes and goats, that span across the Pyrenees in the South, Touraine, and Burgundy, and the Cher, in the center or elsewhere across France, where their cheeses appear on the choicest of all cheese platters, around the world. There are the producers of fruit that have a special label, and peasants who make ham or sausages from the few pigs they raise every year. There are the rooms and the guest houses, and thousands of other local activities that enable people to earn a few dollars to survive. Between the inhabitants of the Beauce region around Chartres, and those who live in the scrubland in the South of France, there exists the same will to survive, but with very different chances of doing so; since the former is subsidized by Brussels, whereas no one looks after the latter.

On second thought, even though they do happen to be true, the descriptions we just gave of the farms do not correspond to any specific one. Whether or not, he or she happens to be dressed like a city dweller, the male or female farmer remains true to form; from generation to generation, they hand down a way of looking at life, of appearances, of looking at things, of family life. They have a specific idea of their work and their destiny, a complex way of looking at things, for many of them, at one and the same time are workers and heads of companies. They all have a deep sense of the weather, and of the passage of time; a deep sense of their independence; an innate pessimism and a sense of fatalism; a profound sense of revolt against a society they feed and they despise. They have a jealous sense of their independence. They depend on subsidies, but they consider them their due, because the market is crazy. This is a due that humiliates them, all the more so because they expect the check, and depend on the bureaucracy that sends it to them.

They have all made immense efforts to modernize, and are disappointed. Two thirds of the farm operations that existed just after WWII, have disappeared: one has to assume that, except for an extraordinary unanticipated change, one quarter or a third of the 680,000 remaining ones will disappear, in short order. There would be much to say about the "long march", the cruel and uncertain march of the farmers of France, Europe, the United States, and elsewhere; about the old farmers who do not have children to succeed them; about the fact that there are many more people leaving the farms than moving to them; about the fact that, in France, the farming milieu is the one with the highest number of suicides, by far. No matter what, at least in France, their collective destiny over the last half-century has been one of the most dramatic of all sectors of society, for farmers have had to change landmarks and relativize their certainties. As Bernard Hervieu states, several breaks have occurred in the agricultural world over the last fifty years: it has become marginalized in a society in which, until very recently, it constituted a majority; carried along by a distant market and stressed by infrastructures that violate their secrets, farmers experience the fact that their impact is diminishing in the territories they formerly dominated; in the past, they provided food directly to family hearths; they are now simply the weak links in networks that transform, prepare, refrigerate what they deliver; in the past, they were subject to the land and the seasons, and worked with their hands; today, they are users of machinery and chemical products that help them violate the laws of nature; wishing to be free, without being able to do anything about it, they are subject to the laws of the marketplace, and those of bureaucrats; they have not forgotten the days when the peasant's word expressed both roots, a state and a trade; they have difficulty accepting themselves as specialists that the modern world forces them to be. But in brief, they have acquired the certainty that the sacred character of the social contract that linked the agricultural world to society, a social contract, that during their last congress the Young Farmers proposed to redefine, has been broken. They are in such a state of disarray that the demands and the capacity for revolt that characterized them in the recent past have vanished.

To show what the cost of progress that has characterized this state of despair has been, we should analyze the performances that have occurred and the results obtained. We will then turn to the state of "our countryside".

Performance and Production

A few figures will give an idea about the extent of performances, as well as the production that took place; of the effort that happened in the agricultural sector, where the rate of productivity has progressed over the last half-century, the most among all sectors of France.

From 1955 to 1999, the population of France increased from 42,8 to 58,5 million people.

In 1945, farmers represented 30% of the population in activity, whereas they represented 5% in the year 2000.

—From 1960 to 2000, the number of farms decreased from 2,080,000 to 663,000. The agricultural land being exploited, diminished by 8% across the country; and the average surface being farmed increased from 14 to 42 hectares.

—The annual average production of a milk cow evolved as follows: 1375 liters in 1961; 2420 in 1971; 3296 in 1981; 4397 in 1991, and 5120 in 2000, but lactations of over 10,000 liters per cow are frequent.

—As for the productivity of labor, two figures will show the evolution: a "laborer" produced 50 bushels of wheat per year in 1900; he can now produce 10,000 today. The number of people "fed" by a farmer has increased from 2.5 in 1945, to 15 in 1970, to 60 and more in 2000.

—The population has increased more than 40% in fifty years, and the standard of living has also increased, while the consumption of food has doubled. Nevertheless, the balance of external exchanges has been turned around, and from an importer, France has become the second exporter in the world of agricultural and food products.

—Whereas between 1960 and 2000, agricultural production has quantitatively more than doubled, its market value has decreased by a fifth in constant francs. In the meantime, the value produced per worker has quadrupled.

—In terms of specialization, the very large farms, that in 1980 represented 13% of the total number in activity, in 2000 represent 30%; during the same period milk farms decreased from 20% to 11% of the number of farms in France; mixed farming from 29 to 21%; whereas, beef cattle farms increased from 6 to 12%; and the "others", the diverse, from 32 to 36%.

—Progress continues with the help of science, technology, selection, organization, but seems to slow down in certain areas and to accelerate in others, through genetics.

At a theoretical level, it can be demonstrated that less than ten thousand agricultural producers could feed all of the French; sixty thousand could feed the fifteen countries of Europe; a million, strategically located, could feed the world inhabited by nine billion people! This leads us to the problem of the non-technological, non-economical limits of performance; for the world is not a production factory, a consumption factory, but a space to live in, while producing and consuming in a nature that is alive.

The Land: Society and the Environment

Rural space is being eaten away and denaturalized. Built spaces have increased by 12% between 1992 and 2000; roads and parking lots by 10% and other spaces (gardens, dumps, quarries) by 17%; in other words, built space has increased more than 600 square kilometers over the last two years. Today, these artificial zones represent more than 10% of the territory of France. At the same time, "useful agricultural surfaces" decreased, while forests increased. Rural demography has been radically modified. At the end of the Nineteenth Century, the rural world had over thirteen million inhabitants; that is to say, it represented 30% of the French population, the same number as in 1960, but it represented only 23% of the total population that had gone up by ten million inhabitants. The aging of the population is evident: in 1955, in agriculture, there were two young people less than 15 years of age, for every older person over 64; in 1993, the young and the old were equal in numbers in spite of many people who have retired or taken early-retirement. The percentage of men on the farms has increased considerably, because women have been the first to leave.

Rural development becomes a process of constructing new forms of social cohesion based on solidarities of proximity, within a territorial framework. The administrative map that was drawn in 1790, and has been modified very little since then, gives way to a new territorial organization into spaces that are not very homogeneous. They are at one and the same time, urban and rural, and combined in such a complex way that our Cartesian mind has difficulty in apprehending this. We must draw up a new map of France that is less inspired by an "abstract" will for equilibrium than by a concern for concrete efficiency and sociability. Reality enjoins us to invent another vision of the necessary coherence between the elements of territory, economy, the socio-cultural and the politico-administrative. Each of the partners, each of the actors, pays a heavy cost for the prudence exercised by our legislators.

Formerly, farmers were masters or serfs on a territory they alone, or almost alone occupied; today they are a minority, or almost, in communes where there exist only agricultural activities. The farmers are frustrated by the riches they create, they feel marginalized in a space that is dominated by the function of production, and that is becoming the refuge of people fed up with living in the city. But at the same time, the city is no longer an enclosed space or an edge. The city is invasive; it is a system of global organization, the heart of multiple networks, equally invasive; it constitutes a place where riches such as culture are created; a heart that will soon have an impact on everyone. This is how rural districts disappear when they are not caught up in an urban dynamics. The city is a hive from which tired bees escape as soon as they can, and impose the tempo of their weekends and their holidays on the rhythm of rural life. City dwellers give country people small jobs, but transform forever a civilization founded on agriculture. It is very difficult to imagine what the future will be.

Before raising specific issues, the progress made over the last forty years must be deeply respected; and it is difficult to be at one and the same time a pessimist and an optimist. Yet, one has to be, since we can neither afford to give up the remarkable results that have occurred in the area of production, nor ignore the real damage inflicted on rural society and the environment, as well as the difficult balances of international markets that must remain open to the destitute. It is by assessing objectively all of the above that we can create tools that will help us to outline the "search for a policy".

Appendices

Fifty Years of Evolution—Statistics and Graphs[1]

A preliminary remark: How can we represent with a few statistics the major evolutions that have transformed forever French agriculture over the last fifty years? Numbers are abstract and arid; comparisons are as difficult to make as units are variable (number of people, various units including constant Francs or not, etc.) and especially when their scale goes from thousands to millions, difficulties increase over time; series explode when they decline or they increase; finally, the number of series and the number of graphs needed to represent all these phenomena would fill a much thicker book than this one, even if it were printed in small print. It is almost impossible for a statistician to establish a homogenous series over a long period of time: even the definition of the agricultural farm varies over time, so when we speak of methods and bases of calculating data in terms of money or it terms of accountability, it is difficult to do so with rigor! We would get lost and also lose our reader if scientific and statistical rigor gave way to the concern to extract and compare major tendencies. I therefore had to simplify, cut, isolate, in order to keep only a small number of the most significant series to illustrate my proposals.

---

[1]The series represented by graphics, are expressed as indices based on 100 in 1960.

1) A Strong Increase in the Volume of Production

The Evolution of Some Areas of Production

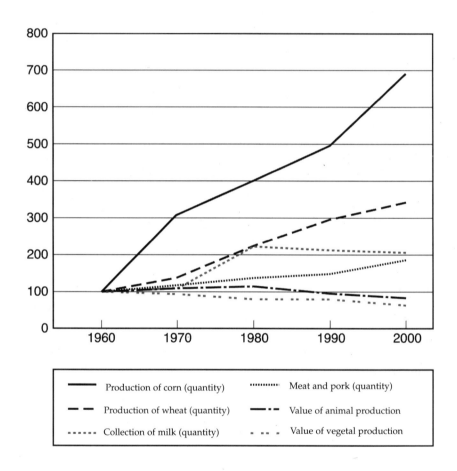

2) ...With Less and Less Work and
For a Less Flattering Economical Result

Technical performance does not translate positively into economic or human terms: in spite of the huge increase in quantity, the value of production in constant Francs continues to diminish over the last fifty years and, if the farmers' income did track that of the average surface of the farms and their production, it's basically because of the massive reduction of the number of producing farms, and of their workers!

Instruments of Production and Economic Results

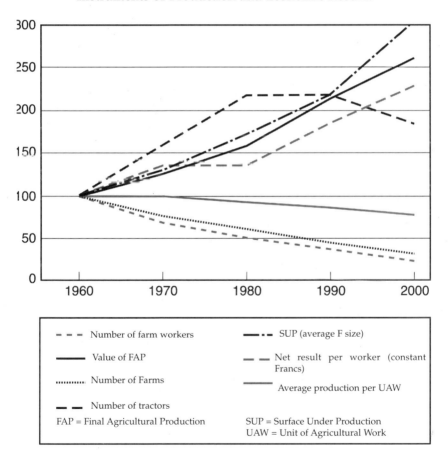

– – – Number of farm workers	—— · — SUP (average F size)	
—— Value of FAP	— — Net result per worker (constant Francs)	
.... Number of Farms	—— Average production per UAW	
— — Number of tractors		
FAP = Final Agricultural Production	SUP = Surface Under Production	
	UAW = Unit of Agricultural Work	

3) Which Leads to Troubling Macro-Economic Indicators

Whereas the number of agricultural and food industries are decreasing in terms of global employment along with the valued added over the last fifty years, the amount of subsidies farmers receive continue to increase until 1992, when the aid system implodes and the reform of the CAP that established a decrease in the cost of production compensated for by direct aid. The various subsidies of the European Union, today account for more than 10% of the value of production and direct aids contribute to more than 50% of those who are eligible, including the most competitive!

A Few Indicators (in %)

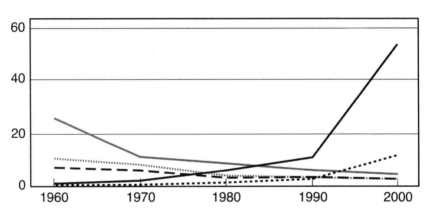

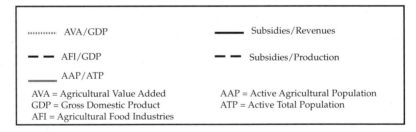

AVA = Agricultural Value Added   AAP = Active Agricultural Population
GDP = Gross Domestic Product   ATP = Active Total Population
AFI = Agricultural Food Industries

4) ... And Few Perspectives in the Future!

The example of the evolution in milk production is especially significant. Whereas some fifty years ago most farms raised a few milk cows
specialization has resulted in a decrease in the number of producers at
a much greater rate than the disappearance of farms. What is more wor-
risome for the future is that in spite of the establishment of a control
over the production of cows' milk by the introduction of individual
quotas since 1985, the average size of herds as well as the average rate
of yield of cows has continued to increase. Without a change in policy,
the reduction of the number of cows and herds seems inevitable.

The Evolution of Milk Production over the Last 50 Years

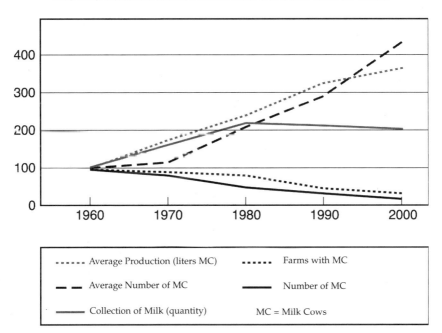

Summary[2]

Farms	1960	1970	1980	1990	2000
SUP (Thousand Hectares)	30 374	29 479	29 479	28 4082	7 586
Number of F (Thousand)	2 082	1 588	1 240	939	663
Average SUP per farm	14	18	24	30	42
Active (Thousand AUW)	3 341	2 464	1 911	1 370	1 015
Number of Active FW per Farm	1,60	1,55	1,54	1,46	1,53
Number of Tractors (Thousand)	680	1 106	1 485	1 476	1 279
Volume of Production	**1960**	**1970**	**1980**	**1990**	**2000**
Wheat (Million Tons)	10,6	13,8	23,4	31,4	35,9
Corn (Million Tons)	2,4	7,4	9,6	11,9	16,5
Oil Producing Plants (Million Bushels)	1,5	6,7	11,8	46,2	56,0
Pork (Million Tons)	1 204	1 308	1 583	1 730	2 305
Cows' Milk (Millions of Liters)	10,8	18,3	23,8	23,0	22,6
Value of Final Agricultural Production	**1960**	**1970**	**1980**	**1990**	**2000**
Value of Animal Production (Millions of F 2000)	188 100	203 472	212 589	176 980	149 812
Value of Plant Production (Millions of F 2000)	339 348	316 566	269 547	268 119	210 202
Total Value of Agri. Prod. Without Subsidies (Millions of F 2000)	544 401	536 649	497 501	459 338	377 739
Value of FAP including Subsidies (Millions of F 2000)	544 401	537 390	498 875	463 958	419 750

SUP = Surface Under Production; F = Farms; AUW Agricultural Unite of Work; FW = Farm Worker; FAP = Final Agricultural Production.

---

[2]The series were established on the basis of a constant 2000 Franc, therefore before the introduction of the Euro (6.5597 F).

Percentage of Plant Productions (%)	62.3	59.0	54.2	58.4	55.6
Average Production/F (F 2000)	261 479	338 406	402 318	494 098	633 107
Average Production/AUW (F 2000)	162 945	218 096	261 054	338 655	431 546
Average Production (Hectare) (F 2000)	17 923	17 969	16 923	16 331	15 216
Economic Results	**1960**	**1970**	**1980**	**1990**	**2000**
Net Revenue of Farms (Thousands of F 2000)	186	177	131	118	98
Net Revenue per Farm Worker (Thousands of F 2000)	69	92	93	129	156
Net Revenue per Farm (Thousands of F 2000)	89	112	106	125	148
Subsidies	**1960**	**1970**	**1980**	**1990**	**2000**
Subsidies for Products (Thousands of F 2000)	0	741	1 373	4 620	42 001
Subsidies per Farm (Thousands of F 2000)	1 460	2 937	6 308	7 783	9 637
Total Subsidies (Thousands of F 2000)	1 460	3 678	7 681	12 403	51 648
Average Subsidy per Farm (F 2000)	701	2 316	6 194	13 208	77 900
Average Subsidy per AUW (F 2000)	437	1 492	4 019	9 053	50 884
Subsidies per Product (Hectare) (F 2000)	0	25	46	162	1 522
Subsidies in Farm Revenue (5)	0,8	2,1	5,8	10,5	52,5
Subsidies/Production	0.3	0,7	1,5	2,7	12,3
Percentage of the Economy (5)	**1960**	**1970**	**1980**	**1990**	**2000**
AVA/GDP	10,5	6,8	4,2	3,4	2,7
VA-AFI/GDP	7,3	6,2	3,4	3,4	3,1
AAP/TAP	25,9	12,9	8,8	5,9	4,6

AVA – Agricultural Value-Added ; VA-AFI = Value-Added-Agro-Food Industry ; AAP = Active Agricultural Population ; TAP = Total Active Population.

Chapter II

"This Can't Continue"

The Assessment Questioned: Product; Space; People

The few figures I presented, demonstrate that production and subsidies "prosper" together. In examining them closely, one notes that in spite of its technological success agricultural development over the last fifty years has not resolved all the problems of the past; whereas it has created other types of problems. Agricultural development has been uneven, and has increased inequalities: farms and regions that are disadvantaged by natural conditions are penalized by the existing system of subsidies; it is as though they wanted to "get rid of them". But what is happening with the regions that are advantaged by subsidies? Pollution, drifting from the land, land concentration, and prices put even these regions in danger. The model that has been developed cannot be sustained. Against all evidence, farmers have by necessity, and institutions have through sheer short-sitedness and stubbornness, continued their practices without taking into account the unpredictable consequences they have on production and the negative consequences these have on all the rest. The model is running out of steam, performances cannot hide the fact that the natural environment has been either abandoned or attacked. Thus, by drifting away from the land, farmers who have contributed to the economic development of the country feel they have been left behind. The decrease in the cost of production brought about by modernization is counterbalanced by the fall in agricultural prices which, for a time, were propped up by agricultural policy. They are now carried along by global market prices that determine everything, even though this makes no sense whatsoever. How can it be otherwise, when agricultural products that are negotiated on the international marketplace represent a few percentage points of the world production? As these products do not represent a significant reality, as they are negotiated as surplus; they are still taken as references, and they determine the price of the entire production. A further perversity of the global system occurs that is the direct result of the choices made when the Common Agricultural Policy was established. The choices which are different in nature have converging effects. The aid given to the agricultural producer is exactly proportional to the volume of production that is contrary to the law of decreasing costs: therefore eighty percent of the subsidies go to twenty percent of the farms that are best placed to produce most cheaply. The second choice that was made is just as serious: by retaining only the basic principal foodstuffs and by excluding those products that were not traded on world markets in 1961, an important part of the European territory and an important number of farmers were excluded from benefiting from the CAP. All of these elements combined have led to the situation we now find ourselves in: market prices, including subsidies, cannot satisfy the most efficient producers of widely consumed food; farmers who concentrate on producing specific foodstuff must make out on their own, and just a small number of them make ends meet only because they invest heavily in quality and professional organizations. No one is happy with this situation. Moreover, everyone is worried because European and international negotiations can make the situation even more precarious. There are undoubtedly some exceptions, though. Let's just say tritely, that they confirm the rule.

Coming back to the 1960s, I would like to remind you of the vision that shaped the organization of the Ministry of Agriculture in France. This vision can be noted in the creation of three general directions related to: 1) production and the markets; 2) space and its development; 3) people and society. Considerable efforts have been made in this area, but the results are disappointing: we are producing more, but prices have come down more quickly than costs; the farmland is better equipped and more charming than ever, but it is experiencing difficulties that threaten its very equilibrium: women, men, children are better educated, they are in better health, but they are abandoning the land or are preparing to do so, for "this is not a life", or rather, "it can't continue", as I have been told time and time again during my travels in the countryside. As a matter of fact, the considerable amount of work done on "structures" has had little effect when market forces are the most dominant and determining ones, par excellence. In agriculture, politics cannot let the market impose its laws, and afterwards correct its effects; it is then too late for anything to change. One must constantly anticipate the market, "negotiate" with it, and intervene. Timely intervention is easier and less costly than belated intervention.

During the 60s, our need, our will to produce more were obvious and the efforts made to produce and to sell were considerable. In quantitative and qualitative terms, the results were surprising, and yet the agricultural world and "global society" can't help but feel a sense of failure that we still have to analyze. We have to find an answer to this state of disenchantment and anxiety.

Fears of Progress, Social Demands, the Origin of Opinion

A dynamics of opinion has arisen in opposition to firms, markets, technologies and science; also farmers. Still uncertain, often pertinent, sometimes lively, unjust and almost violent, this opinion expresses four

major concerns: the healthiness of food; risks related to genetic manipulation; the environment, and the decline of rural life. Let's examine
each of these concerns, before focusing on the creation of this dynamics
along with the hope accompanying it.

Concern for the safety of food is greater than ever. Mad Cow
Disease and hoof-and-mouth disease do not constitute an authentic
social risk. For the most part, the place they occupy in the public eye
results from the emphasis politics and journalists give them. Be that as
it may, a form of doubt has invaded the collective conscience and the
most minor incident can trigger it. Although the specter of famines disappeared a long time ago in France, fears can crop up at any given time.
The most serious fear is related to genetic manipulation, plant and animal combined: the former concerns the environment, and the latter,
medicine and ethics. As far as plants are concerned, opinion is split, but
the majority of the population demands that we take our time and
count on politics to stake out a clear position and decide what it will
simply adopt. In the minds of the public, animal genetics and transgenetics conjure up genetic manipulations that could eventually affect
human subjects. The problem is not related to experimentation, but is
situated at the level of principles; we do not have time to solve the quasi
religious ethical problem raised by cloning. This respect for life that is
asserting itself today, vaguely has the same origins as the obsession
with the environment, a return to nature that expresses itself with such
great passion because nature itself is threatened. In addition, there is the
preoccupation of city dwellers who demand that rural life retain its
authenticity and its soothing calm; that it remain a refuge for those who
are disenchanted with urban life.

Except for militants, none of the above preoccupations contribute in
a decisive way to the constitution of public opinion. But this is how the
fertile humus of a new civilization finds another expression in the success of peasants' markets and farm products. These fears, these tastes,
these tendencies become requirements; demands become a political and
commercial given. Industry and commerce that have done so much for
the standardization of food products become the promoters of French
products, and the institutions continue to set new standards; associations proliferate to prevent the development of a quarry, the construction of a highway or work in a factory with smoking chimneys. All this
is too diffuse, as it is at one and the same time categorical in its refusals
and its claims, uncertain or excessive in its demands, but all this sets off
major changes that will develop, insofar as it also expresses a certain
fear of progress. And this fear of progress insidiously will become a fear
of all progress. This is how a break occurs between an information-
based civilization, where all forms of risk taking are permitted, and a
civilization structured around the quality of life and daily living where

citizens seem frightened by their own audacity and now seek refuge in the daily. This is how the effort made to facilitate the emergence of a new wisdom founded on the respect for life—the foundational value and cornerstone of modernity—can find its legitimacy and democratic support. Let's hope that this wisdom does not simply reproduce a form of pietism from another era. Will we ever understand that our destiny is not to be found in a "return to nature", no more than in a shattering of its laws? Our destiny lies in cohabitation.

The Vicissitudes of Common Agricultural Policy

The CAP was a remarkable political audacity; it was indeed a wager on the solidarity of countries that were at war in the recent past. It was the fascinating adventure of a continent whose unification was given as an example to the world, and an exceptional chance for the agriculture of the six founding countries to realize its potential. However, the CAP evolved into a proliferating bureaucracy, a teller's cage, where everyone got used to asking for more, a boxing ring where a grab-bag of interests fight it out with one and other, a shrinking budget with increasing obligations; a green carpet around which an increasing number of countries negotiate concessions, instead of working out a common vision to be negotiated globally and concretely, at one and the same time. Europe has become a springboard for globalization, much more that a melting pot for identity. The reason for this is that over time, after a period of illusions, two opposite visions about the construction of Europe confronted one another. Agriculture was the only common major European Community Policy, but it soon became the scapegoat of a Union that had doubts about itself; whereas farmers were the only state prosecutors calling for a political space, unable to say what it was or what it wanted to be; unable to administer a domain which, forever and everywhere, had been the primary preoccupation of human communities.

I would like to try and explain how this came about, in a few bold brushstrokes. In so doing, I would like to downplay an indictment filled with bitterness about what was done, and about the fear of the future that people experienced.

In Search of a "European Agricultural Model"

Beginning with the 1970s, environment and rural development began to be integrated within the CAP, when a policy was introduced to deal with mountainous and disadvantaged regions. This was the first time that the CAP dealt in a positive way with areas that were not linked to agricultural activities. In 1985, a "Green Book" established this: "We perceive more and more that the role of agriculture in a modern industrialized economy is not only to assure economic and social strategic functions, but also to conserve the rural environment". In 1985, the Unique European Act gives a true constitutional basis to European policy on the environment, and environmental directives follow. Nonetheless, while maintaining the same orientations the reform project of 1992 favors intensive producers and encourages concentration and specialization in the most heavily subsidized areas of production. In 1995, a group of nine experts is given the responsibility of analyzing the problems and inconsistencies of the CAP, and of defining the principles of an integrated rural policy. The report points out that the means of intervention the CAP has are obsolete; and suggests a "Common Agricultural and European Policy for Europe". In November 1996, the European Commission organizes in Cork (Ireland) the European Conference on Rural Development, whose task is to define "a European agricultural model", along with a coherent and lasting policy for rural development. This will be a failure: a number of states, and in particular France, do not want the CAP to adopt a broader policy. One would have thought that with "Mad Cow Disease", the "new" social demands and the current increasing membership of the Union, this error would have been corrected. Indeed, while affirming that rural development is the second pillar of the CAP, "Agenda 2000" emphasizes the fact that European agricultural policy is subject to the rules of the World Trade Organization! This is followed by the Berlin and Göteborg Summits that, at one and the same time, affirm their adherence to the market and to the rest, but without stating how. As the financial means were not put in place, there was a definite contradiction between the ambitions asserted, and the instruments provided. No answer is given to the haunting question: how, with a constant agricultural budget, can the European Union deal at the same time with an increase in production, an increase and a diversification of tasks, an increase in the number of member countries, whose agricultural population often represents more than a third of the entire population? With increased membership, the Union's agricultural lands increase by 38%, and the number of its units of production, by 75%. The Union also decides to devote only 25% of its budget to these newcomers. The Union has neither the means to carry out its policies, nor the policies related to the means it can devote to it. Unconcerned about their own contradictions, the oracles and experts should understand and wonder why no one is following them on this.

The Quest for a Compromise

As it was very aware of the impasse its agricultural policy had led to, and as it wanted to comply with the World Trade Organization's injunctions, the European Commission adopted to abolish economic, social and political links to any assistance; and subsequently came up with the idea of a subsidy per hectare. This was another mistake for: 1) farmers do not want social assistance; 2) the abolition of economic, social and political links, just like the subsidy per hectare, contradicts the idea itself of a rural economy and an agricultural policy, because they deprive the politic of a tool to regulate production, and discourage the "market organizations" that we could, on the contrary, expect a great deal from; 3) the system favors large farms, and leaves the "peasant" regions that represent more than half of the European land mass without substantial aid; the "specific" productions that hold a real promise for the future were also left without aid; 4) it in no way favored the non-productive activities or the public good that Europe claims to support; 5) it will orient the agriculture of Eastern Europe in the direction of huge properties, without any consideration for the peasant populations that actually make up the immense majority of the farms.

Thus, the CAP is nearing the moment of truth, and the European Union is being asked to define itself. The situation can be described in four phases: 1) every political entity is responsible for ensuring safe food for its population, along with the safety of the environment; if it does not, it abdicates its responsibilities; 2) food and environmental policies are the flip side of agricultural policies; 3) the policies invented in 1961, and reformed since, are no longer relevant; they cannot last because they do not take into account new realities; 4) having reached a point of crisis for other reasons, the European Union would lose all geopolitical relevance if it did not assert the demands of its civilization and if it did not propose reforms in the political arena on which the future of agriculture and the food-supply of the world depends.

Agriculture, Food-Supply, Environment, Farming, World Perspectives

Over the last fifty years, the number of human beings that are insufficiently fed has not increased at the same pace as the population of the world. This is a remarkable result. But should we be satisfied with this? As the population is "likely" to increase, can these same results that are satisfying from the point of view of an overall percentage rate, be maintained over the next half-century? Is there reason to hope that, progressively, and in spite of the increase in population, the number of under-fed people in the world will in any way decrease? Will we have to wait for the population of the world to stabilize, or decrease, for the world to be able to feed all its living human beings? The somewhat optimistic answers that we often hear are founded for the most part on the diffusion of technology, but they do not take into account three essential variables: the evolution of the objective elements of production; the consequences that modernization has on the "peasant population"; the financial and organizational capacities of the countries where peasants are most badly fed. Two extreme cases enable us to gain an insight into the "agriculture; food-supply; environment and farming perspectives" of the world: the example of Africa that I shall analyze in greater detail, and the example of China that will be subject to a more lapidary overview. These choices cannot be contested. They will be analyzed, after I develop a few general considerations. All studies in this domain point to the incredible needs of a world whose population will increase considerably over the next half-century. Now, of the six billion people "alive" today, one billion are barely "surviving", or slowly dying; in the near future, nine billion people will have to be fed. Therefore, production will have to double to feed the population of the world. We can surmise that the needs of the population will triple because eating habits that evolve will dictate this, especially with respect to meat consumption that is the great consumer of calories! Let's turn to the demographic hypothesis and say why "we must" feed every human being, and who especially can make this happen. The best case imaginable would be for people to feed themselves, and I would like to suggest the ways and means, along with the conditions necessary to make this happen.

As I noted, without taking a great risk, we can estimate that the population of the world will be close to nine billion souls in fifty years time. AIDS can slow the process down, it can, alas! handicap tens of millions of people, especially women, but without rendering them sterile. Lower birth rates in a number of countries, for example Egypt and Tunisia suggest an evolution, but its effects will take place later on, after the middle of the twenty-first century. These countries need help. Only the education of young women, and family planning, can change things. The former is happening slowly, more slowly than in the past, for the educational system world-wide is in full crisis; the latter is contested by believers who want to "let life have its way", even if it means condemning millions and millions of newborn children to hunger.

Abortion comes up against beliefs and traditions. In order to understand, one has to go back far into the past where, because they didn't have pensions, parents had to have eight or ten children in order to "keep" two or three to ensure their old age. This was dictated by the state of medicine of the times. Making necessity a virtue, the "reputation" of a woman was measured by the number of children she had. Pity the poor women who had no children! These prejudices still exist today. A slow evolution is taking place, but before we see any results Sub-Saharan Africa will have a population that is double, or almost double today's. Young women have to be educated against prejudices and beliefs; their own prejudices and beliefs will have to be modified, and they have to be taught practices that save them and save society, along with their own descendents, from undesired births. We have to contest Washington's political decision not to contribute to help finance a competent international organization for birth control; we have to oppose the Vatican's doctrine for whom the problem of hunger will be solved by demanding the right to life for human beings who have not yet been conceived, and are already destined to starve to death.

But no matter how hostile we are to the "right to life", we have to do everything to ensure that each child born, even in the most destitute family in the poorest country, has the right to live sheltered from need. I am not pleading here for an end to poverty, but for an end to misery, because poverty can be resolved in the near future, whereas it will take a hard struggle to bring an end to misery. We have to take up the struggle. We owe it to ourselves, for the motto of the French Republic promotes the right to life, above and beyond political liberty and civil equality. But political reason and even our very own future exist above and beyond ethics. For, between the people of good will we happen to be, and a handful of "infamous terrorists", there exist an immense mass of resigned individuals who are now impatiently awaiting better days. Are we deaf and blind; do we not feel the earth trembling under foot; have we not understood that billions of miserable people are actually waiting for a cataclysm to bring an end to the reign of the white man's world? And, doesn't the economic sphere exist, above and beyond politics? Even though the spasms of the "Papal Bulls" have subsided; it is still the case that reduced to a billion privileged individuals, the world economy will go through crises, because these people live for the satisfaction of invented needs. It is time we understood that our prosperity now depends on the development of a world market of essential goods. The poor are the future of the world.

Africa was self-sufficient a few decades ago. It is less and less so today and, as its agriculture is breaking down, it is becoming more and more dependant on other agricultures. Because it is obliged to import food, it does not have enough currency to import tools, machines, fertilizers, to modernize and progressively become autonomous. The mines, factories and plantations that are entrenched in camps, and are cut off from human and economic life-blood, have created few jobs and given a few fistfuls of money to some governments, but they have not favored development. It is about time agriculture became rehabilitated as the mother of all balanced and stable development. Contrary to the arguments put forth by "modern" developers, far from endangering the future in question, agriculture is the future. There are four reasons for this. I have just raised the first one: by importing grain, sugar and milk, Africa does not import machinery; it does not build the networks it needs; it does not educate its populations. The second reason is due to the fact that, in order to develop, agriculture needs the help of craft industry, of a whole series of cottage industries, of diversified commercial networks that all together will create a thrifty middle class able to found new institutions. The third reason is obvious: sixty five percent of the population of Sub-Saharan Africa is rural. Hunger is rampant in the bush, and each year hundreds of thousands of young people migrate to city slums, thereby creating a dangerous imbalance and insoluble problems. Only a developing agriculture and the multiple activities it generates can give life to a rural world, and encourage the creation of small towns and villages that can organize this immense space, and create proper conditions for life. Agriculture is not part of the past: it ensures the present and can, if we are careful, create the conditions for a true form of modernization. The fourth reason is related to the uncertainty around an essential point: can the world feed nine billion people; can it in the long-term without Africa and the Third World Countries have enough arable land to ensure everyone's daily bread? We cannot take a chance on this. Faced with the problems that are now raised regarding the environment, no person of responsibility has the right to consider that the full development of the production of African agriculture is useless. The poor countries and regions need their agricultures, and the world needs all the agricultures of the planet.

But is this possible? Are the dreams of agriculture and the end of hunger simply illusions? We can answer these questions by having recourse to the absurd; do there exist other ways to foster development; are there other ways to prevent the spread of local conflicts that threaten world peace? Agricultural development does not oppose mining development, the installation of factories to transform raw goods, mechanical factories and information technology, nor factories capable of transforming and exporting products from the land. It can very well adapt to all of the above, but only agriculture can ensure what all the other projects taken together cannot: the stabilization of rural populations and the development of socio-economic fibers that found the prosperity and "governability" of countries. The problem therefore is not related to the relevance of the agricultural solution, but to the capacity of Africa to define and establish a real agricultural policy. The current state of confusion is such that Africa cannot do this alone.

Without wanting to take the place of Africans, I would like to enumerate a few simple ideas that could be implemented in a first phase over a ten year period: 1) Through import duties, protect African agricultural production from external competition with their artificial prices; 2) Compensate for the costs resulting from this for the populations concerned, through an international contribution; 3) Establish systems of mini-credits that would enable each peasant to acquire the necessary fertilizers and tools; 4) Finance the stocking and the protection of harvests; 5) Sign contracts with all the "plantations", encouraging them to become centers of agricultural popularization and progress; 6)

Provide food subsidies which, when food is sold at national prices, would create resources totally dedicated to agricultural development; 7) Within the framework of concerted programs, give African producers special quotas of exports; 8) Give local radio stations the means to broadcast daily, and at specific times, programs that popularize agriculture that would begin with ancestral practices and would guide them to adopt more efficient practices; 9) Give mixed national governmental organizations the task of creating a center for popularization, help and activities in each "natural region"; 10) In each major African region, create a "Center for the Observation and Orientation of Agricultural Development", made up of African and international economists, sociologists, agronomists, teachers, but also of real peasants who come from other parts of the world. Together they would prepare ways of reaching the second phase.

This second phase would consolidate what was accomplished in the first, and would lead to the definition of all the types of investments that could accelerate modernization and increase production, all the while respecting the environment. It would enable the world to discover that Africa can feed itself. It needs to, and we have to help the continent do so. These simple ideas are contrary to current principles and actual practices. They all stem from an idea that is even simpler; let's give back to African farmers the confidence they have lost in their land, their climate, their own know-how, and make them responsible for their own future. Anyone who has had the first hand experience of observing African farmers knows how hard working and clever they are. They know that it is not by handing over machines and transmitting practices from abroad that they will reacquire a desire to produce, but it is only by returning to an agriculture they know will they be able to advance.

The debate on the development of Africa has always been tainted by resentment by certain people, and contempt by others. Recent history makes it all the more difficult, as the performances of Indian and Chinese agriculture give the West the impression that nothing more can be done in the South Sahara, and because the situation facing the continent makes Africans believe that after having been mocked, they feel they are now being rejected. This is catastrophic, and a dialogue must be renewed on very different bases. After their long history, after slavery, colonization, improvised independences, it is not surprising that the continent is experiencing a series of flare ups. Let's turn to references from a different source; a book that was published some fifty years ago, based on serious arguments, predicted that Brazil was the Eldorado of the future. It still is. The authors of the study ignored that growth is not a mechanical phenomenon, that democracy is not an institutional system that can be bought and sold. To succeed, democracy must be founded on development, and then must be invented and constructed by those who want to live it. During the 1960s, after so many centuries of external domination and methodically organized haemorrhaging, Africa did not have the capacity for self-governance, especially since it was hemmed in within borders imposed by colonization. With the help of illusion, Africa had some success, but because the continent was not aware of its own weaknesses, it soon reverted to the practices it had been taught within the framework of Sovereign Statehood, especially corruption and favoritism. We have to dare to speak up, and denounce all doctrinal attitudes and all useless rancour. Over time, Africa must be able to govern itself, and contribute to the riches of the world. We need this to happen, and Africa needs the help of our riches and our skills.

But what's the use quibbling, when everything seems to condemn Africa? Globalization that isolates it from all strong currents, investors' preferences for promising investments, the development of subsidies given to farmers by the United States and the European Union, the decision taken by Washington to no longer finance family planning, the interest in terrorism that is rampant in other countries, the fear of AIDS and the blaming of Africa for it, the detestable reputation, and the persistent help the major powers give to leaders who abuse human rights, Europe's political inconsistency that, needless to say, renews the Lomé Convention, but is unable to define and adopt a real policy when dealing with a Continent that in a certain way it is responsible for, and whose future depends on this. Is this enough? Undoubtedly not, for the despondency of the poor; their lack of any real future is, in part, the result of the wastefulness of the rich. What good is it to plead in favor of the poor, when we very well know that the resources of the planet are limited, and that it is impossible for nine billion people to eat and live to their fill, if the wealthiest among them do not put their model of consumption into question? The buying and selling of pollution rights, the rejection of ecological practices, the waste of energy and water the rich are guilty of, in fact, condemn the poorest of the poor. The only solution is agriculture. But both the Africans in power, and the continent's intellectuals, must accept the fact that this is their last chance and that, if they do succeed, they will take their place in a world that neither awaits nor rejects them.

One can imagine that China's problems are very different. I must tackle them, though, even if I have not experienced them first hand. I will base my analyses on common knowledge, and will refer to the work of Claude Aubert, a scientist at the National Institute for Agronomical Research because his studies have enabled me to measure the size and the complexity, the long tradition and the great ambition of a country I am ashamed I know little about, first hand. There are 775 million rural inhabitants, of which 325 million are in activity. Many regions have the population density of Holland or Belgium. Agriculture occupies 65% of the population, and represents only 15% of the GNP. There are great differences between the plains in Eastern China, and the mountains of the West and North. In addition, there exist relative urban affluence and rural frugality. Famines no longer occur, and the great majority of the people eat their fill and consume a great deal of pork and poultry. They produce honorable yields, and real progress has been made through the selection of seed and the increasing use of fertilizer. There are few industrial farm producers and factories turning out farm machinery. In this country, there are important seasonal migrations and continuous movement to the cities. On the whole, today the country is self-sufficient from the point of view of agriculture, but it sometimes reverts to imports. There are two long-term major challenges: the limitation and, even scarcity, of water resources that are absolutely necessary for agriculture and, in every domain, a real need for investments that defies the imagination. Those in power hardly analyze these issues, and do not say how it will be possible to address them, all the while demonstrating a confident belief they will be.

The most serious challenge comes from the immense and frugal population that endures less and less its status and its low income; a population that is truly worried about the future. China now knows that modernization cannot ensure the survival of small family farms where six or eight people, helped by two or three workers, live on and live off 6,000 square meters of land. Public authorities are hence trying to imagine how, and at what pace, the country can absorb and support such massive rural exodus. Migrations are no longer forbidden, as in the past; cities are given help to increase their capacity to integrate and employ those who are flocking there, but the problems remain, for the drop in price of agricultural products encourages migration to the city. As a matter of fact, after having been "administered" then "liberated", without a doubt agriculture needs to be "assisted". The question remains, but how? If it happened to be impossible to do so, then how could cities deal with migrations? Claude Aubert concludes: "it is at a very critical juncture that choices have to be made, favoring agricultural revenues or the dynamism of sectors of the economy that are open to the external world". In its traditional wisdom, China is attempting to distribute the enormous pressure that the great mass of its impatient "peasants" represents. The country is modernizing farms and attempting to create a rural life that will decrease the "need to leave"; it is undertaking major works of all sorts, and developing industry and exports. China is not threatened by poverty, but rather by the enormous problems of an ambitious adolescent full of life. An adolescent with a rich culture and great know how. This can certainly be seen in the incredible development of certain products: electronics, appliances, textiles and leathers that increasingly occupy a large percentage of the world market, because of the competitive edge it has, thanks to the networks and know how of a labor force that is more skilled than demanding. All of this leads us to conclude that China is attempting to solve the problem of migration to cities through a strategy of external investment and world trade. The process is so important and the masses in question are so numerous, that no one can remain indifferent to this today: because of China, but also India, Pakistan, Indonesia, Latin America, the world could be inundated by computers and refrigerators, but be short of food!

To be totally relevant, this study of world-wide agriculture food, environment and farming perspectives should not have been limited to Sub-Saharan Africa and China: above and beyond the so-called developing countries, it should have taken into account Russia, whose capacities transcend difficulties. It is nonetheless true that this brief survey helps us discover a number of certainties: the world is capable of feeding nine billion humans under certain conditions; to do this, though, requires a huge effort. Above and beyond ideological quarrels, it can and must define a policy whose aim is to ensure that each person has a daily supply of food, that each peasant has the desire, the possibility, and the right to produce what he or she needs in order to live. If this does not happen, there is a great danger of creating a humanly unbearable situation.

The Unacceptable "Possible Futures"

Now, more enlightened, let's return to France, to study our own "internal" perspectives. Created by the DATAR, established with the help of the National Institute for Agronomical Research, and directed by Philippe Lacombe, the Group "Agriculture and Territories", in 2002 published a remarkable study entitled, "Possible Futures". This document sets out four scenarios that are complimentary more than contradictory. I would like to give a brief summary of them.

The first scenario is inscribed in a logic of continuity; it suggests that European protectionism follows international rules. This is reassuring since it closely follows what already exists; but the changes envisaged to comply with the WTO may deepen the contradictions that exist today between economic functions and social functions. It proposes to reinforce the quotas on production, as well as the sanitary and environmental rules, but with a certain European Union preference to favor the juxtaposition of two complementary agricultures: "competitive" agriculture for the world market; and "territorialized environmental" agriculture for the production of goods and services—including the private sector—targeting aware and demanding consumers.

This scenario takes into account the difficult cohabitation of the two forms of agricultural development, the two types of territories, the two "mono-functional" agricultures. The second scenario, that is industrial, is based on free trade with a progressive disappearance of government intervention. Business strategy is the main driving force, here. In spite of resistance from consumers and certain farmers, this scenario favors the pursuit of intensive technologies, the widespread use of GMOs and of hormones that can ensure the competitiveness of our agriculture. The two scenarios presuppose world growth and a liberal Europe.

The two following scenarios both explore a new social contract that promotes regions, territories and rural development associated with local groups. In the first of these two scenarios, agricultural producers continue to exercise activities that ensure primary production, and are encouraged to participate in secondary and tertiary activities of trans-formation and commercialization of their products. This agricultural policy stresses quality as its major preoccupation, and it is not con-cerned with standard products. The balance between the two agricul-tures is fragile, and conflicts of interest can be predicted. As far as the final scenario is concerned, that of service agriculture, it opens perspec-tives for a new rural economy where agriculture co-exists with other activities, whether or not they happen to be linked. Not very credible in the short-term, this scenario is more long-term as it can correspond to a postmodern socio-economic and political context.

The exercise that was initiated by the DATAR is very useful because it rehabilitates the future, the art of possible futures. But it soon reaches its limits as it spells out tendencies rather than ends; it traces the conse-quences of imaginable choices. It does not define a reality: "...whether or not we happen to favor or oppose internationalization, French or European agriculture can no longer be analyzed in isolation: every deci-sion concerning it must take into account the external world". This is how Philippe Lacombe defines the constraints he has imposed on his group, but without it being said in so many words, at the same time we are led to believe that each of the scenarios imagined has something that is unacceptable within this framework.

The Indispensable "Impossible" Future

We are now back to this bureaucratic and diplomatic refrain with a thousand couplets on unsatisfied needs and the dangers in the world, but a single refrain: between "we must" and "we cannot". "We cannot change anything. We can do nothing about it because, until we get unanimous agreement, we must maintain the rules of the game, because after the failure of an administered economy, we are left with only the market, because if the South doesn't know how to feed itself, we can develop new sciences and technologies to help it meet its needs; we can help the South, but on condition that this does not diminish our standard of living. The impatience expressed in the world does not express the cries of the famished, who receive only derisive answers: all we need are a few subsidies to appease this agricultural discontent; the problem of the environment is a false one, the demographic explosion is in the nature of things, and is proof of the miracles of medicine, the two and one half billion farmers who want to emigrate to the cities will stay where they are..." Well, No! Instead of imprisoning ourselves in the same international rules and agreements, we should start from the needs of human beings and of the world; let's re-examine our script. The market is a mechanism, hunger is a need. Let's start from the latter; then say how the former can satisfy it. Agricultural policy is a tool, and the safety of foodstuffs is a need. Let's begin with the latter, and then say how the former must be redefined. Everywhere, farmers are necessary to ensure our subsistence and our territories, the adaptation of their practices and their farms to the demands of the market is a necessity; rather than blindly making practices subservient to the market, let's see what we can do to transform rural spaces, over the long term. Fertilizers, pesticides are helpers, nature is a living reality. Let's make certain that use of the former does not put nature in danger. Let's change fertilizers and pesticides, and not nature!

We should not lock ourselves into the single logic of the indefinite growth of production: we should accept, practice, call for a dialogue, where the human and natural costs of this growth are compared with the profits it generates. If, in spite of all; if certain people believe that current international rules and agreements can feed the human species, safeguard nature and lead to peace, we must refuse to share in their illusions or, rather, in their ideology but at the same time, we should also propose a plausible alternative. If we do not believe; if realities prohibit us from believing that a green nature, an end to hunger, the life of our societies, peace can automatically flow from the free market, and the unbridled forcing of the laws of nature, if; however, we do not believe that the market is the primary regulating mechanism, let's try and reach a pact, a compromise, a concerted interaction that will not paralyze scientific and market dynamics, but will help us avoid sacrificing the fundamental values and realities that these dynamics must serve.

In brief, I would like to affirm three certainties:

1) The Common Good concerns both human beings and nature; it is long-term;

2) Science and the market are dynamics that contribute to this; but alone, they can become uncertain, and even dangerous masters;

3) Only through the mediation of a concerted policy can society assume this contradiction.

Chapter III

In Search of a Policy

Europe Is Not the Exception

It is not easy to give a clear description of the world-wide-food-producing industry. Nevertheless, I would like to attempt to do so:

- In Africa the population is about to double in the next thirty years, while production is stagnant, and up until now no true development has been defined with its agents and its dynamism. Poverty that was restricted to the countryside has invaded cities that cannot cope. It is urgent to find a solution to this problem.

- In China, India, Latin America and Indonesia, the population is increasing, self-sufficiency is not at all ensured, modernization cannot take place without structural policies that need great investments. Together, these countries and continents today represent a population of two and a half billion people; and in thirty years, some four billion inhabitants. It will be extremely difficult for them to be self-sufficient; still more difficult to say what will become of the hundreds of thousands of peasants forced to leave their plots of land. It is certain that none of these countries can hand over the responsibility of solving their serious problems to the market.

- Eastern Europe remains an unknown quantity: It possesses lands that are often excellent and after long and difficult times agriculture in this immense region is being reinvented. The states intervene, oscillating between great rural landholdings inherited from the past and more modern-scale structures that would pave the way for a large numbers of modern peasants. With few exceptions, in the end a specific culture and economic necessities will determine the outcome in favor of large landholdings; especially because demographic pressures are less pressing here than in other regions. Public efforts, major investments and time will be needed for this to happen.

- In the United States and Canada, large, mechanized, modern farms are tightly linked to important industrial and commercial structures. In these countries, farms are going through difficult times, and since they cannot survive on world prices, farmers are abandoning the Mid-West or Alberta, where family farms are doomed to oblivion. Washington has established massive public interventions, both to prop up production and to safeguard natural and social space. Ottawa is currently considering what to do.

- The Cairns Group adopts an all conquering free trade position based on exceptional national or social conditions. And surprisingly, part of the developing world is adopting this position against its own self-interest.

- Then, for forty years the European Union alone has dared to declare that it has established a policy of supporting agricultural prices, but it seems to apologize for doing so.

This overview gives rise to a number of questions:

— Can we be certain that the disappearance of all production subsidies would alleviate competition waged by agricultures from developed nations against developing nations; and that, without these subsidies, developing countries would progress towards self-sufficiency?

— By adopting policies based on the principle of free trade, do we not run the risk of developing a "system" where a few tens of thousands of large concentrated farms linked to major firms in industry and commerce, and located in a few privileged regions in the world, would marginalize the production of hundreds of millions of poor peasants distributed around the world who would experience greater poverty while their countries' chances of further development would disappear, while the food safety of the world would be ensured by profit margins?

— Can the "Western model", that globalization tends to present as the one that can be applied generally, be suitable for all the agricultures of the world? Can we run the risk of making a large number of agricultures disappear, while all countries need a form of agriculture, and the world needs all its different agricultures?

— Notwithstanding, can we be certain that the disappearance of financial aid, which represents a serious risk for the majority of the agricultural firms of developed countries, would not lead to the invention by these countries of thousands of schemes to intervene at other levels?

— In short, rather than focusing on the single problem of free exchange alone, would it not be better for the world to approach the problem of production, food-supply, rural population and environment, globally and planet-wide? Instead of beginning with a principle and applying it to the limits of the absurd, would it not be wiser to look for a synthesis that can evolve over time, and that would take into account the globalization of exchanges on the one hand, and the diversity of knowledge and needs, the merits of the markets and the demands of nature, of subsistence and societies, on the other?

Europe is not the exception, and there is only one exception: the Cairns group that brings together all the others, Australia, Argentina and New Zealand: for objective but exceptional reasons, perhaps even provisional ones, the countries that make up this group can export their products without subsidies or protection. We should keep this in mind, but we cannot act as though the entire world were composed of large landholdings on the pampas, with a work force that can be shrunk or extended and deployed at will, unless we happen to be dealing with new and limitless spaces, whose resources have not been exhausted by centuries of human occupation.

Why intervene and what is a policy?

We therefore need agricultural policies so that everyone can be fed and that the world can feed the world as it should. Let me explain why:

— Since agriculture is responsible for the most obvious aspect of our safety, the Common Good demands that it be given the means to ensure this role. In spite of what is claimed, it is false to say that we can count on world markets to ensure our subsistence. They may have a surplus today, but they certainly won't tomorrow. Market prices that can be down today may not be tomorrow, because it is not certain that the disappearance of subsidies will not set off an increase in world-wide prices. The proof is in the past, and an increase in demand confirms this hypothesis.

— Fluctuations in market prices weaken the systems of production that are characterized by their "inertia". You don't close down a farm the way you close down a factory, because you won't open it up again.

— World-wide agricultural market prices have a tendency to decrease, because they are fixed on surplus markets. World trade on which one wants to fix prices, represents only a few percent of world-wide production and consumption: except in a few exceptional regions of the world, no well-managed farms constantly produce food at cost.

— Agriculture is one of the victims of a "civilizing" evolution, whereby the modern world remunerates less and less the people it needs more and more: farmers, teachers, general practitioners, nurses...

— Our taste, health and "quality" requirements are increasing constantly; and they can be satisfied only within the framework of agricultural and food-supply policies.

— "Man does not live by bread alone". In addition to food, agriculture gives society the services it needs. Why discourage farmers, and force them to abandon the land, when all this will lead to is replacing them with guardians of natural space?

In brief, each country or region needs to create policies relating to matters of food-supply, national and regional development, and the environment, as well as maintain the ability to negotiate its future in a world that is becoming more and more uncertain. Now, the time is ripe to think about and define a new attitude because: - discontent that puts productivism in question is spreading, - increasing membership in the European Union raises issues for agricultures, both among current members and among future ones; -the United States is putting in place more extensive measures than we are in Europe; - the awakening of a political conscience and the strategic importance of the imbalances of the world are such that world-wide geopolitics is, and will be, more and more influenced by farm-produce considerations. The time and the need have come for the WTO to establish an institutional regulation of agricultural markets in all their diversity, product by product, region by region, moment by moment. It is therefore better to define the concept of "agricultural policy", than to object to it. The difficulty of doing this is enormous. And, this is because there are so many divergent interests, contradictory expectations, ideological positions that oppose one and other. Because this problem occurs at every level, it is impossible for us to think of "France alone"; "Europe versus the United States"; the "generalization of the Western model"; "hunger in the world" or "developing world storming the developed continents". We have to think of all of these together, and we will now try and say how.

No matter what "subject" we are dealing with, "a" policy is made up of a number of elements that we can outline theoretically: 1) an awareness and critical analysis of a need, of a possibility, of a new problem, of a crisis, a contradiction, a fundamental imbalance; 2) positioning and evaluation of the importance of the element in the general problematics; 3) identification and open consultation with the "agents", and the "victims"; 4) evaluation of the "cost" of non-intervention; 5) determination and articulation of the values, the objectives, the criteria and disciplines that define this policy; 6) synthesis, negotiation, mediation and decision; 7) establishment of an entity delegated with "responsibility" for the whole; 8) determination and mobilization of the necessary means; 9) establishment of a system of periodical evaluation and revision.

It would be bold and cruel to compare the vicissitudes of European agricultural policy to this methodological outline.

From the Simple to the Complex: New Paradigms

As we progress, we can see that the concept of agricultural policy is becoming more complex. Once we get beyond the problems of production and subsistence, it becomes more difficult to define, to explain, to govern. Agricultural policy has to play the role of mediator between different, and sometimes contradictory, needs. It was easy to talk about production subsidies when ration coupons still existed, or were still in mind. We did not have to explain the reason why the French Nation was asked to do so, but simply justify the choice of techniques to make this happen. It was not difficult to justify market intervention, since it did not cost the taxpayer anything, and was paid for by consumers who were too happy to find in stores, all they had gone without during the war. Everything has changed: basic needs are taken care of, except for a minority that we have unfortunately gotten used to ignoring more or less; taste, the dietary quality of food, and its ease of preparation, concern the consumer more than the price itself because, as we have said, the percentage of family budgets devoted to food has continued to decrease; a need for nature and existential anguish are such that the threatened environment and the "French desert" have become significant themes in the political arena; the rise of the Third World; the widespread spectacle of the poverty of hundreds of millions of humans, and the illnesses that accompany this poverty, now contribute to a form of pessimism that challenges the optimism technological progress has instilled in us. Humanity must structure the ideas it has of the world, while its needs, its work, its demands, its aggressions, its boldness continue to make it more and more complex, more sensitive and more fragile, more difficult to manage.

Let us sum up this policy in terms of a few principles. To be brief, they were presented in the "Introduction" of this book as "themes" to be reflected upon. In light of our analyses, they become the paradigms that can organize the thinking of institutions, citizens, city-dwellers, as well as farmers, and of all of society.

—1) "The question of subsistence is too vital, too poignant to be strictly a social question" (Jean Jaurès), or to be strictly a question of economics. It calls for a political solution, as does everything that ensures safety.

—2) Even if it could, the Western world cannot have the ambition of feeding nine billion humans in 2050. The world needs all the agricultures of the world, and each country has the right to feed itself. The poorest ones cannot exist otherwise. For the world to find its balances, it needs to invent, in the North, a productive agriculture that respects nature and favors maintaining authentic rural societies and, in the South, a modern peasantry that can feed and retain the masses that cities, industry and public services cannot look after.

—3) The struggle against poverty and hunger, in the world and in each of our countries, does not exclude charity but is a matter of justice, and also of the principle of prudence.

—4) The Western World has a model of consumption and a standard of living that are wasteful. If these were to become wide-spread, all of the natural resources of the world would be exhausted in less than one generation. The Western World could live better, without wasting.

—5) Since world prices are depressed and variable, the European Union and the United States hand out subsidies to farmers that often exceed half of their revenues. This cannot continue, and it would be proper for consumers to pay a price for their subsistence that corresponds to actual costs. The European Union must achieve this, and allow its excess production to be exported, but without competing with agricultural products of developing countries. The Union must favor the development of production that corresponds to its energy needs and industrial fibers.

—6) Agriculture must respect nature. Farmers must look after it. Production must no longer be their only activity. They are now responsible for doing work that contributes to the general good, and that affects the environment and the life of rural society.

—7) Medical research linked to agronomical research has made important progress possible in the areas of health, food production and reproduction. Today, they have reached a point where our natural heritages, and sometimes our convictions are threatened. It is neither up to scientists nor the market to determine where we are going. It is only up to society to make these decisions, enlightened by scientists.

—8) Wisdom has it that societies around the world, and the agricultural world in each national or regional entity, should reach an agreement that ensures the former has the food safety and the environment it needs, and the latter, the long-term equilibrium necessary for it to fulfill its obligations.

An International Agreement: Globalization or Local Governance

Do the vision and the policies we have just sketched out have a chance of being considered at the international level? Surely not in the past, perhaps not today, but certainly tomorrow!

However, we can no longer go to Geneva and be judged by the World Trade Organization with its one dimensional market values. The market is a good mechanism, but cannot be accepted as the only law. Institutions are the guardians of treaties, and cannot be laboratories where the rules that give them power are cooked up. It is now time to put things into question because the world is not at all the promised Eldorado, and the United States is applying policies identical to the ones they have always opposed; because the future of agriculture and peasants in China and India; the future of food production in Africa, raise problems that the current rules cannot solve; because conflicts are arising around genetic breakthroughs, and serious environmental risks that are fundamental. In the recent past it was affirmed and even recognized that "people had the right to feed themselves"; today, following the same logic, we must define and recognize the right of people to ensure their future according to their own values, in their own natural environment.

Yet, nothing can happen by simply putting things into question, we need to present an alternative, but before doing so, it has to be developed. It will be developed, neither by the WTO that has a tendency to reproduce itself; nor by the United States whose administration, for political reasons obviously, is the champion of globalization and cares very little about environmental risks; nor by Argentina, New Zealand or Australia who, for social or geographic reasons, think that their future lies in winning over global markets, and that no precautionary measures, no social evolution should stand in their way; nor by China or India who, after having made enormous progress, have not taken the exact measure of the social and political upheavals caused by the modernization of their agriculture; nor by Africa because the unconditional freeing up of trade will in the very short-term lead to cheap food-supplies, and this continent hopes to feed itself cheaply at the risk of not being able to renew its agriculture; nor by Russia where without a doubt there exists one of the greatest, but one of the most uncertain capacities in the world to cultivate and distribute food-supplies, and which still must get over its bitter agricultural adventure. Nor again by the institutions of Europe, unless the Union decides to enter the debate at the level of its geography, its climates, its land, its civilizations, its interests, and not from the perspective of an exclusively driven market ideology that it will soon fall prey to.

Before reforming the Common Agricultural Policy, or rather before defining its new Food Production and Distribution Rural and Environmental Policy, the expanded European Union must become aware of its own responsibilities. The Union has to know that a new model, that only it can come up with, at the crossroads of the laws of the market and the demands of the territories, can enable the world to feed itself, based on the development of all the agricultures of the world. Any policy that would not take this route would be dangerous for the balances of the world, because it would be oppressive for the traditional agricultures of Asia, Latin America, Indonesia, Europe that today represent half the population of the world, and soon, three fifths.

It is not a question of abandoning the idea of defining the rules or the necessary dynamics of the market. Instead, the diversity of natural environments, civilizations, populations, and their needs, must be respected, and the principles of the governance of the world must be founded on this diversity. Founded on rules that are imposed, since not respecting diversity could lead to insubordination and revolt; whereas, on the contrary, respecting diversity can be an invitation to seek collectively a world-wide Common Good everyone senses a need for and a commitment to, but no one is willing to take the political risk of defining its underlying principles.

How can we not make a plea for international organizations finally to devote as much attention to satisfy every person's needs as they do to the globalization of the market? How can we pursue this policy, when it leads to an exacerbation of the problems that weigh down the world? Let's look at three examples: it is unacceptable for the OCDE to consider water as a market commodity, and choose simply to ignore those who cannot afford this scarce resource; it is totally unacceptable for the WTO to approve the ready distribution of genetically modified organisms long before they have been proven harmless; it is scandalous that a market has been created where the major polluters acquire the rights to pollute, when they themselves are assailed by pollutions that threaten them. Beyond these examples, chosen among so many others, the essential thing to remember is the principle that underpins international trade: the country that produces the cheapest, cannot expect to export endlessly and compete with the agriculture of less developed nations that, in turn, cannot compete with the agriculture of developed countries. The country that produces the most cheaply, cannot expect to conquer markets that are territorially defined and that aspire to ensure the safety of their supplies. Who cares about those statisticians, for whom the world is a world market, and all one has to do is show that in twenty five years time the market will be faced with a global supply that can statistically ensure feeding one billion human beings. Statistically!

These facts and ideas are beginning to firm up. Other facts inform us that the poor, or their cousins, can set up obstacles on the star-studded roads of the triumphant powers. Adversaries of globalization and adversaries of the political world-order have become accomplices; by inventing another policy founded on an agreement one can propose to the world governance through diversity, as an alternative to globalization that prevails today, and makes everything uniform. Europe can do this, because it does not have the attributes of a State, and it is going in the direction of creating a new type of political entity (a type that has not yet been identified), a civilization that resembles no other, a balanced and powerful economic equilibrium. It wants to build on its diversity, in each of these areas.

A European, Agricultural, Food-Supply, Rural and Environmental Policy

The word governance can take on very different meanings. This is the case with world governance, and European governance of Agriculture: through consensus, the former defines the codes of behavior; whereas through deliberation, the latter puts in place positive measures of intervention. This explains the difference between an international "agreement" and a European "pact". Starting from the paradigms retained, we now have to determine for which interventions the European Union can and must take responsibility, while respecting the rule retained by international governance. Whether they happen to be budgetary or statutory, these interventions are predicated on the member governments exercising discipline, for whom agriculture remains mainly a domestic matter.

The Europe that grew out of the Community was a Utopia filled with promise. The Europe that will grow out of the Union has not yet taken shape. The Europe of the Euro is a daring idea, but no one can know where it will lead. The redefinition of decision making institutions is an endeavor that can uncover more difficulties than it will resolve. In its current state that has been fuelled by expansion, Common Agricultural Policy can facilitate the emergence of both a geopolitical vision, and a real solidarity. In the past, it symbolized and embodied a certain Common Good. It has been weakened through various defensive reorganizations undertaken to counter, without actually facing them, problems that have arisen because of the development of internal production and the evolution of world markets. It is still a reality. But it is threatened by difficulties encountered, and by conflicting, different and even contradictory conceptions of the global and internal; the commercial and farm-produce industry. To be a presence in the world, the European Union must have greater ambitions than those of its founders, different also from those that undoubtedly suit other countries and other continents. This is an urgent matter.

This is timely. Knowing that if there is collective agreement on a real agricultural, food production rural and environmental policy, if it is proposed while demonstrating that this policy would not endanger agricultures that are changing or developing; the Union could count on support from China, the Indian Sub-continent, Sub-Saharan Africa, as well as Indonesia and a significant part of Latin America, where there are serious production, agricultural society and environmental problems.

But it would be absurd to define a policy by beginning simply with the farmers' needs. It is imperative that these needs, and those of society, respond to one another. This is the meaning of the appeal launched by the Young Farmers in favor of a pact with "Society". How can we not respond to them? But let's attempt to understand the exact meaning of the undertaking and the proposal. Both follow the general logic of modern societies that in all matters have a tendency to substitute a contract for a statute: it is by following such a logic that the young farmers called for a new order of the markets, the lands and territories that took into account different values and interests, while evolving with them. The new order can be founded only on a balanced exchange of agreements that we can outline as follows:

Farmers guaranty society sufficient food of a specific quality; they commit to producing economically the elements of this subsistence, all the while respecting nature; they acknowledge they are the principle guardians of rural space as a living and organized reality.

Society guarantees that farmers will receive internal prices corresponding to the cost of production of good-sized and well-run farms. The cost of surplus exports must be born by producers. The tasks and responsibilities exercised by farmers, beyond production of food stuffs, are to be remunerated as "services for the public good".

The European Union will define and negotiate, with international organizations, the rules that apply to the internal market, as well as those that apply to exports destined for external markets. It should develop a policy to combat hunger, so that aid favors agricultural and food-producing development of the countries benefiting from it. It should define the principles, and keep a watchful eye on the consistency of national and regional interventions, in matters of services related to non-production; and finance these interventions, taking into account national and regional realities.

No matter how transparent these reciprocal agreements happen to be, they cannot take the form of a classical legal contract. This gives rise to the idea of a quasi-contract, by which each party helps the other; their potential difficulties are settled by mediation, rather than arbitration or juridical means. It can only be a question of a pact entered into in good faith, even if the actual violation of the rules that govern all parties can be the object of sanctions, or measures of retaliation. But let's not dwell on the formal elements of this quasi-contract, and let's get to the point: this treaty establishes a method of defining, negotiating and governing that is democratic, and can permit both parties, the agricultural world and society, to understand that each one's life depends on the other; and having understood this, each has to guarantee it for the other.

Part III

A European, Agricultural

Food-Supply, Rural

Environmental

Policy

A Proposal

1: Strategies and Elaboration of the Policy

Preamble

—Water, food and energy are indispensable for life. A lack of these vital elements has been at the root of past and present armed conflicts. The safety of food-supplies is the condition for a lasting peace. Because of its old civilizations, its democratic cultures, its sharing of the Mediterranean, its geographical proximity to countries of the former Soviet Empire; because of the Middle East, of the Maghreb, of Sub-Saharan Africa; because of its actual international economic importance, Europe can play a major role in the construction of peace that depends on geopolitical, geo-economical and geo-strategic worldwide balances. It is within these parameters, that Europe must develop the agricultural policy it needs.

  Its "nourishing function" remains agriculture's fundamental responsibility. It requires technological, juridical and economic measures that involve the responsibility of the Political side of the State.

—The subsistence of every human being is not ensured today; and will it be tomorrow? Serious doubts remain about our ability to ensure adequate food-supplies tomorrow, for the world population continues to increase; whereas urbanization, desertification and pollution, decrease the possibilities of production.

—Scientific and technological progress that have made possible and will continue to make possible increases in food production, have undergone developments that call for ethical and ecological inquiry.

—Today, more than half of the world's population farms. Half of this half experiences poverty and hunger. This distress, the workings of the market and the unequal and discriminatory modernization of agriculture, induce mass migrations; while the cities cannot indefinitely accommodate those who seek refuge there.

—The current agricultural practices of the poor countries, and those of industrialized ones, do not guarantee the protection of the natural environment. On the contrary, they often threaten it.

—Mass demonstrations that are increasing in developed as well as in developing countries reveal the existence of public opinion that is raising questions about the future of the species, and even about that of the Planet. The International Community must find a solution to this: 1) by respecting the Unity of the World and the Diversity of the

115

natural, social and cultural environments; 2) by taking into account the long-term, and respecting the principle of prudence.

—At the very moment it is preparing to reform its now antiquated agricultural policy, the European Union, that is a new kind of world power, a determining commercial power and an original agricultural and regional civilization, cannot limit this reform only to problems of self-sufficiency and export strategies to the satisfaction of its immediate agricultural and rural needs and their book cost. There can be payout only if local, national, European, and world agricultural, food-supply, peasant, rural and environmental variables and uncertainties are examined.

—Founded on a multi-disciplinary approach to the complexity of the issues, this new policy must satisfy the different demands made on it. There cannot be two policies: one devoted to production and the other to local and general balances; one dealing with the immediate and the other the long-term. One driven by an uncontrolled dynamics, the other adjusting its pace to the rhythm of nature and society. There can be only a single policy that ensures the mediation between sometimes contradictory, but equally striking or demanding needs.

—The Union must define itself, in order to tackle this problem; managing diversity in unity and the present in the long-term can be done only through common institutions that articulate a vision and a policy.

—After having defined the needs of global society, the Union must mobilize the means needed for farmers to be able to address these issues more fully. This is why a Pact to establish a lasting relation between the whole of society and its agricultural domain was outlined. The vital functions that agriculture is responsible for justifies it. This definition is a founding political act that requires debate and adhesion.

—At the very moment it is defining its policy and taking into account its increasing membership, the European Union must state its positions on the worldwide need for food, the right of people to feed themselves, the environment, the life of soils, the quality of food, the control of practices and innovative products.

Proposals:

—Following the example of what was done at Stresa forty years ago when the agricultural clauses of the Treaty of Rome were implemented, and following the example of what is done today in Brussels with respect to Institutions, prompted by the European Commission, the European Council and Parliament should organize a Conference that will define the "Agricultural, Food-Supply, Environmental and Rural Policy", (EAFERP), Europe needs. The Conference should situate these needs within a demographic, human, environmental, as well as economical and commercial worldwide perspective. New members of the Union would take part in it.

—Presented as a coherent quasi-contract that once formalized by the Union would involve European society and its agricultural component, the Conference would define the principles, institutions and practices of this policy. Subject to democratic and participatory institutional procedures in place, and adopted before January 1, 2005, this Policy would be applied progressively before January 1, 2012, fifty years after the birth of the Common Agricultural Policy, (CAP).

II: Content and Management of the Policy

Preamble

—In all the member countries of the European Union, in all the countries that are being integrated into the Union, and in all the countries of the world that have reached a certain level of development, the cost of food in family budgets, and the cost of agricultural products in the price of food products on the market have not stopped decreasing over the last fifty years

—In the same countries, a social need relative to the quality of food and agricultural practices that respect the environment and even the well being of animals have been evolving rapidly and asserting themselves. This social need calls for a significant redefinition of the methods and systems of production as conditions that are necessary for society to sustain its agriculture.

—For certain essential foods, no farms exist either in the United States or the Soviet Union that can cover costs of production by selling their products on today's market. Subsidies often represent more than half a farm's revenues, and yet for no plausible reason, certain food products receive no subsidies whatsoever. If there do exist in the world a few countries where costs and prices meet, this is because of exceptional circumstances, or questionable social practices. One must take these into account but it would be dangerous to fall into line with such countries.

—The rural exodus continues in spite of subsidies. It contributes to unemployment, because the model founded on a parallelism between industrial development and the modernization of agriculture today is outdated. It cannot be a reference point for China, India, Africa, Latin America, Indonesia, where together the farm population represents two billion people.

—To convince the greatest number of rural inhabitants possible not to migrate, a complex policy must be worked out, where the regulation of the market is accompanied by interventions in the rural environment, and many farmers take part in it once it has become multi-active and multi-functional. This will give rise to a modern rural civilization.

—Needing to produce more and more, having experienced the harm brought on by practices concerned exclusively with increased production, troubled by the situation of rural populations of developing countries, the Union must undertake research and encourage policies favoring economical and efficient agricultural practices that respect nature.

—Addressing both food-supply and agriculture, the European Agriculture, Food-Supply, Rural, Environment Policy (EAFREP) must take into account the fact that, between agricultural production and consumption, activities related to collecting, preparation and distribution are being developed that will have a determining influence on the evolution of agriculture and on the relations of farmers to the markets. This influence is creating worries, because industrial and commercial firms are concentrated, whereas farmers are legion; also because, in order to control their supplies, using contracts or direct controls, these firms are beginning to integrate various sectors and are weakening the farmers' ability to negotiate. European policy must ensure that a code of behavior be established that respects the autonomy of the different actors.

—Because of scientific and technological progress, the productivity of labor and the production per hectare or per animal have not stopped increasing over the last forty years. But the dynamics of techno-science leads to innovations that make it difficult to accept progress and, in the short and long-term, raise serious ethical problems for civilization that society alone can resolve.

—Malnutrition, demographic growth, the evolution of food practices to satisfy all human needs make it necessary for world food production to be multiplied by two-and-one-half to three times, over the next three decades. This target can be met only if: 1) "the right for people to feed themselves" is respected and upheld; 2) food subsidies from the outside, instead of competing against the development of the beneficiary's national agricultural production, favor it; 3) once their role in feeding the world is recognized, farmers everywhere will be able to live from the fruits of their investments and their labor.

—The goal of self-sufficient food-supplies cannot be attained by eight or nine billion people who expect to consume at the same level Western societies do; it also cannot be attained if Western countries themselves do not accept to put into question their wasteful models of production and consumption. Everything forces us to adopt values, organizations and practices that help us find satisfaction and comfort in saving our available resources.

—The market alone cannot deliver this result as it has a tendency to amplify imbalances and tensions. For competition to thrive, it must constantly be regulated, and it must also be accompanied by other measures since a number of essential needs cannot be met either by its mechanisms or its dynamics. These needs must be defined from a social perspective and met by means of adequate mechanisms.

—Competition can play out in a meaningful way only if agricultures that happen to be at a comparable level of development compete with one another. The opposite case can lead to the massive disappearance of the means of production that are necessary for the collective safety of food-supplies, which particularly depends upon protecting and supporting agricultures within coherent geographical spaces, with a low level of investment and organization.

—But new needs are cropping up that are related to the environment, rural society, the territory, as well as the qualitative safety of food-supplies. These needs are expressed by people around the world who are not satisfied with the mechanisms of the market, and they bear costs that have to be met by public budgets. They also create work that has to be dealt with as a matter of priority by farmers, along with the entire rural population.

—New doubts and a certain fear of progress are appearing that give publicly funded research important responsibilities because, as the guarantor of the safety of discoveries, it must be able to march to the tune of all research undertaken.

—Because of the importance of the problems raised and the consequences the solutions adopted can have for global society and the environment, as well as for farmers and rural inhabitants, it is imperative to work out a global policy in the form of a global agreement between the agro-rural world and the industrial and commercial complex.

—Farmers are as diverse as landscapes, geological relief, climates and histories fashioned by nature. On the pretext of favoring exchange, it would be cruel and dangerous to subject all of them to the same rules. Units producing basic foodstuffs for extensive markets must coexist with specific products that bear a local controlled label. Farms that are market driven and favor performance over rural society and the environment, and farms trying to survive, along with society and soils, must be able to exist along with production. For them to coexist, they have to be given equal attention.

Proposals on Agricultural and Food Production:

= refusal to "uncouple" subsidies and the generalization of "bonuses given per acre". A decisive step towards a destructive integration of the European market into the world market, the uncoupling would compromise the food safety that Europe has just recently attained. This would be profoundly unjust.

= definition of regulatory norms making up the basic code of best practices that can be applied to the farm-produce industry. Public intervention would depend on whether or not the norms are being adhered to.

= fixing European internal prices calculated on the costs of production for modern, efficient units that respect the best practices in question.

= the establishment of an internal system of food subsidies for the poor.

= fixing quotas of production that, when exceeded, each specialized sector will assume the responsibility for exporting them outside the European Union.

= recognition and respect of the rights of countries, or regional economic organizations, to protect their domestic markets against destabilizing effects of the erratic fluctuation of market prices.

= conclusion of agreements defining evolving strategies that facilitate food subsidies today is necessary, just as is supporting the agricultural development of countries or regions that are experiencing serious food deficits. Concluding preferential interregional agreements with neighboring regions.

= establishment of mechanisms or interventions that make it possible to respect the concern expressed in the Treaty of Rome on family enterprises, and the conditions under which rural activity and life can be stimulated.

= negotiation and definition of contractual "codes of conduct" that protect farmers from being vertically integrated by companies. Support for the professional organization of unions of producers that have the real capacity to impose discipline and to negotiate.

= in matters of intervention, the definition of general homogeneous ends and practices adapted to the diversity of production, soils and, provisionally, of the situation of new members.

= accompaniment of initiatives undertaken in matters of quality and labeling.

= along the lines of what exists in the United States, the establishment of a "European Commission of Certification of Agricultural Practices and Food-Supplies" endowed with investigative means and power of decision that would make it possible to ensure the safety of agricultural practices and of these food put out on the market that should correspond to and respect labels.

= development of a system of insurances, linking individual initiatives to public counter insurance in matters of natural and non-natural risks.

= support for a methodical policy that attempted to develop non-food agricultural products progressively and on a grand scale in order to prevent exports from paralyzing the development of agriculture in poor countries, and ensuring that all of our means of production are exploited.

=reinforcement of programs of agricultural support for countries that are candidates for membership in the European Union.

Proposals on Agronomy, the Rural and Environmental Worlds:

= a series of policies related to soils and territories must be added to measures related to production, products and marketing.

= to be implemented, these policies need local and regional, public or private structures. They should be autonomous, and dedicated to defining and organizing the multiple and various tasks that farmers, under contract, can take on. These tasks must benefit from regional, national and European financing.

= particularly in difficult regions, these policies favor economic activities that are complementary, or external to agricultural production, proper.

= diverse from region to region, it is desirable for a modern rural civilization to persist, and all the while continue developing. While influenced by them, this civilization should not be dominated by cities, whose ambitions continue to be asserted in opposition to the power of the State and the farm heritage. Having agriculture as its major constituent element, this rural civilization can better exploit the natural resources related to water and forests. It is necessary to help the farm heritage affirm itself. A just balance would be facilitated by a new definition of "local", "rural-urban" communities. Modeled on reality, they would transform the urban and rural components of our society.

= rural society must develop activities that create jobs and resources. Whether or not they happen to be linked to agriculture, they need investments, and must be accompanied by education. They also point to the need to reexamine the fiscal and social status of moral or physical persons who settle in a rural environment.

= the diversity of rural activities, and taking into account work that is not directly devoted to production, suggests that agricultural colleges and high schools programs should be reformed and diversified.

= as far as its organization and implementation are concerned this policy in question can be only national or regional. The European Union defines its principles and rules. The Union must facilitate studies and research, regional activities and initiatives, exchanges between regions, and help handicapped regions.

= respect for the environment and restoring it, are two conditions for ensuring rural vitality. This cannot be limited to the definition and the protection of nature parks and protected zones, since all activities, and particularly agriculture, must respect the environment.

= the Territorial Farm Contracts falling under the Law of 1999, were destined to favor rural development. New definitions, guaranteeing that they are enforced in accordance with their intents, would enable them to play a privileged role in relation to the proposed reforms.

III. Negotiation and Time-Frame of the EAFREP

= There must be a single European Agricultural, Food-Supply, Rural and Environmental Policy; its principals must be coherent and its administration must be decentralized; its implementation must be diversified to take into account the diversity of concrete situations; it must be applied while respecting the principle of subsidiarity. It can be modified if it is analyzed methodically, and all its results and consequences evaluated.

= Reexamined systematically at a given date, using the same type of process as when it was established, by taking into account the multiple internal and external evolutions that affect it, the policy can be updated without being derailed.

= Involve the policy in international negotiations.

-that the market tends to make uniform, what nature and history have created as diverse,

-that have a tendency to consider only production and the product as determinant factors, to the detriment of the environment, societies and, sometimes, public health,

-that, under the pretext of temporary surpluses, and without any consideration either for rampant hunger or the increasing population, runs the risk of making agricultures disappear. These are essential to feeding the populations of the world, and for the survival of numerous countries that rely on a meager form of agriculture to live and develop,

-that are not sensitive to world events, and seem to refuse to take into account the agonizing and threatening cries of those left behind,

=European policy is being proposed and supported, not in the form of a refusal of the demands of negotiation, but as a modern concept, of worldwide common interest, of agricultural activities considered in terms of the goals of subsistence and food safety, of safeguarding nature, of giving vitality to rural spaces, and confirming the public interest of agriculture and the eminent responsibility of the Political.

Part IV

Messages
to Those Responsible

Presentation

The evolution that French agriculture has undergone and managed, undergone or managed has been breathtaking. From an eternal order and enclosed fields to a time that spans the surface of the planet without borders, the agricultural world has gone from the harmonious alliance of the sacred, the social, the economical and the political, where everything could be explained by its function of nourishing people, to a movement that is not content simply to put into question its values and its functions, but is endlessly redesigning its natural landscapes, and drastically changing its mental ones. The agricultural world is now responsible for safeguarding the social balances of a rural society that it does not inhabit alone, and the environment that covers four-fifths of the surface of the country. This is too much, and nonetheless farmers must willingly take on this responsibility, on condition they be given the means to do so. This is how the idea of a pact with "global society" came about that would give farmers the means to live and to progress and, thanks to the market, cover their costs of production, their non-commercial activities of public interest, being remunerated and honored in a different way than they are today. Do all the actors agree with this idea?

But who are the actors? Beyond the various elements that make up the French community, there is Europe, the World, the former, because it directs agricultural policy, the latter, because of free trade and "globalization, because of problems that arise related to food-supply and peasants. Obviously, farmers are among these actors, but they no longer live in isolation; there are the scientists we expect miracles from, and whose discoveries we fear; there are the industrialists and merchants, who are the prime movers of the market; there are the citizens, consumers, people and citizens, who have become used to feeding themselves at prices that do not cover costs, and who have become aware of the "needs of nature". Finally, there is a planet in turmoil: a human species that is wasteful in certain parts of the world, and poverty-stricken in others. When all is said and done; over the last sixty years the balance sheet of modernity has not been positive enough around the globe for us to continue to forge ahead along the same path. We find tragic insufficiencies and threatening excesses, along with considerable progress. Hence, the time has come to begin anew from the perspective of needs: those of people and society, those of nature, those of knowledge, those of the economy; those of Peace threatened by inequalities, but also those of the farmers of the World.

With whom should I tackle and discuss these very serious and interdependent problems; where can I bring together, at one and the same time, all the actors who are the accomplices, or victims of a machine that "stubbornly" takes the wrong direction? Not having found a single tribunal, and knowing it is impossible to say everything in a single statement, in my mind and separately, I chose to confront representative audiences that expressed the diversity of the points of view and interests I identified above. In front of each audience, I chose to discuss the problems that concern them. The challenge for me was not to tell everyone what their problems were, but to imagine remarks that would each address a specific public, but taken together would transmit a coherent vision. I will therefore present five "messages to those responsible". It is not a question of a literary or a rhetorical exercise but it is a formidable test, because I had to dialogue with five different cultures, five different logics, five different or contradictory interests, while remaining faithful to the ideas that drive me.

The most indispensable of all publics is difficult to unite since it is none other than society itself and, as it is everywhere, it just happens to be nowhere. Yet, as it is everything, it is both diverse and contradictory. I am addressing an assembly of citizens. My remarks contain several messages: - as the agricultural product is a market product, its production imposes restraints that society must take into account; - it is wonderful, but dangerous over time, for us to pay less for our subsistence than it actually costs; - it is perverse to exercise policies, where the budget encourages production that attempts to conquer markets, under unfair conditions; - it is therefore necessary that fair costs cover the real costs of production; - rural society alone can ensure this, and it has to be adequately remunerated; - above and beyond trade exchanges, we have to be concerned with the global economy.

In Brussels, I tell the members of the Agricultural Commission of the European Parliament that Europe is betraying its soul when it forgets the shrinking size of its natural spaces, the history of its landscapes that have been slowly shaped by human hands, the balance that it patiently is trying to build between the market and the rest; when it is tempted to forget that safety of food-supplies is the first obligation of the Political. I try and convince them that the reservations globalization provokes in our old Europe do not isolate us from the world: we are not alone in trying to find a form of agriculture that is both sensitive and market driven. In addressing European parliamentarians, I try and demonstrate that the next reform of Common Agricultural Policy cannot follow the current direction it has taken, and that the political future of the Union depends on judicious agricultural reform.

Within the framework of the Organization of the United Nations, in New York, I claim that, in order to meet the needs of all humans who will inhabit the planet in the next quarter century, we will have to increase world food production by two-and-one-half, or three times. This is not impossible, but it is not certain we can get there, either. We have to give ourselves the means, or openly declare that we accept that hundreds of millions of people will experience dire poverty and hunger. The United Nations must become aware that they will have to face problems of safety; for ending hunger is the precondition for Peace.

I address the assembly of scientists more solemnly than I do the other groups. Though the scientific community has greatly contributed to the progress we can see around us and that we benefit from, human genetics enables the best in medicine to happen, but the worst in humanity. I attempt to convince them that, whether it happens to be private or public, discovery is a common good that no one has the power to use without the consent of human society because, in case of error, along with nature, society would be the first to suffer. But while private research is bolting ahead, public research must be alert, and create instruments that permit wise decisions to be made.

Speaking to Young Farmers, I try and show them that the function of nourishing the country is theirs. They must organize their productive labor; work on the land, and their social life, beginning with this function in agreement with, and with the support of society at large.

These "Messages", taken together, are an "account of the purpose" of the European, Agricultural, Food-Supply, Rural and Environmental Policy set out above.

Paris

To an Assembly of Citizens

For what reasons, in the very heart of Paris, did I organize a meeting on agriculture and, especially, why does the room where we are meeting seem too small? My presence here is less of a surprise than yours. Life has made me a person concerned with giving agriculture the place it deserves in a society that is becoming more and more urban, and whose wealth of goods makes it indifferent to problems related to daily food-supply. As for you, ladies and gentlemen, you are both concerned with making ends meet, and citizens concerned with the Common Good; as taxpayers you are worried about the costs of agriculture, you are friends of nature; in brief, you are city dwellers. For these exchanges to bear fruit, I propose to examine the problem from two points of view; the concrete and the political. As a militant in favor of a global approach, I link together agriculture, production, the safety and quality of our food-supplies, the environment, finally the rural world that inhabits and breathes life into four-fifths of our territory. This global approach, that certain people refuse, is the only one that can ensure the actual satisfaction of the needs I have just listed; it is only a real concern for the environment that can guarantee the preservation of a long-term productive agriculture, along with a vibrant rural space; it is only the autonomy of our subsistence that can ensure our safety and the health and taste of our food; remuneration of agricultural production by the market, is necessary for rational economic development; remuneration of non-food producing work that farmers accomplish to safeguard the natural environment and rural space by contract, is the only way to make this happen; we can ensure our long-term survival, only by close-ly interlinking all of these components.

My assertions are not founded on presuppositions, but on long and careful observations of French, European and World realities. These assertions are acts of faith, founded on analyses and projections, that make it possible to confront a dominating and conquering market ide-ology, that can drive European policy along a path that leads to the loss of our agriculture, our quantitative and qualitative food safety, along with that of natural space and a rural society that are threatened by the separation of food production and territory. Before analyzing those issues that concern you most directly, I would like to tackle globaliza-tion, not because it actively frees up exchange, but because it is a delib-erate policy of standardizing production and consumption, and because it will end up delivering the world into the hands of the most

129

powerful and best placed producers and merchants. The world is politically diverse; and this is so for cultural reasons that have an historical origin; for historical reasons that are founded in nature that itself is diverse. The wish to see Europe become the exact copy of North America, and to see China, India and Africa modernize themselves along the lines of the Western World ignores completely the very nature of these countries, along with the agricultural spaces and rural societies that are necessary to guarantee the balance of the world's food-supply. Biologically, culturally and economically, we need the diversity that nature and history have provided; the world needs it for the sake of subsistence, because in order to feed sufficiently a prolific human species, it must rely on all the agricultures of the world. To plead in favor of diversity is not to condemn unity, but it is to affirm that unity and diversity complement each other; they are necessary for the equilibrium of every milieu, whether it happens to be natural, social, or political.

Building on these first approaches, I would like to formulate four proposals: 1) competition is a good mechanism, the market is a good dynamics, but it cannot be our only law; 2) there is a fundamental incompatibility between the diffusion of the Western World's model of production, consumption, development, and the claim to reduce the untenable inequalities that exist in the world; 3) the European Union cannot afford to be without an agricultural policy, no more than can any country, taken individually, or any other political entity, because only such a policy can ensure its safety; 4) as it cannot build the future of its agriculture on subsidies, but only on prices that cover the costs of production, the Union must force the consumers that you are to accept an increase in the cost of food. But let's continue.

The Western World is the richest, but also the most wasteful entity on the planet. I am aware that these two facts are closely linked but if in the near future nine billion humans each consumed as much energy as the inhabitants of developed countries do today, all the world's natural resources would be exhausted in ten or fifteen years time. If we are not careful, the production of the biomass could turn out to be insufficient. In spite of scientific progress, the world's food-supply needs that are increasing at an impressive rate, will force mankind to cultivate all of the agricultural land available, especially, insofar as a certain portion of the most productive lands will be invaded by water, the desert or urban development. The scarcity of water available for human use, and for irrigation, is already a serious problem. Yet, if it wanted to, the human species could feed all the individuals on the planet, but on condition that they consume wisely, and that they respect nature and invest boldly. But nothing is certain here, and we must question radically our wasteful model of consumption. It is possible to do so in every domain. This was proven in the area of hydrocarbons in the seventies; a policy of conservation of energy proved we could all live well by consuming less. But the inhabitants of the Western World are not alone on this planet, and no matter what scientific progress we make, our voracious appetite for space, water, energy, food, make it impossible to reconcile our ways and standard of living with those of people who live in misery, or only in poverty.

It still remains that the Third-World is evolving; it is making demands to evolve in order to experience the safety and comfort we do. We therefore have to consider another aspect of the problem: increased production of farm produce has brought about a considerable decrease in the number of farms. Drift from the land, has freed up a work force needed by industry. Growth in urban jobs and a decrease in agricultural employment have been complementary aspects of economic development; however, they are a model that most of the world cannot follow. The Third-World itself has a rural population of more than two billion people, who live badly on small parcels of land. Modernization along Western lines would bring about rural drift that no one could deal with. We need to come up with a modern farm agriculture that is productive, without devastating the job market.

The above forces us to re-examine our ways of thinking and behaving, and it is only after doing so that we should envisage our future. Population increase and unsatisfied food needs, a lack of energy and water, the diversity and fragility of nature, an increase of non-productive spaces, the overabundance of rural populations in the Third-World, the will to feed all human beings sufficiently and the need to develop all the agricultures of the world, make us ask ourselves how we can consume less for the world to live better. This leads us to put into question the statement made by President George Bush Senior, that his citizens would never accept a lowering of their standard of living. But above and beyond the questioning of the former President's statement, I want to affirm my vision of the future, and sketch the agricultural policy the European Union needs. It is not by slashing this policy that we will be able to come to grips with the future but rather, it is only by defining it in a specific way that we will be able to meet our long-term needs without losing sight of those of other people. We have to act against our wasteful ways, by adopting a model of consumption that is more economical, but does not prevent us from having the kind of life we want. But first, we need a policy that ensures the safety of our food-supplies.

I will limit my remarks to problem of the imbalances of food-supplies in the world, and I shall undertake this analysis by referring to, with his authorization, Laurence Roudart's[1] "Essay on the Future",

---

[1]Associate Professor at the Institut National Agronomique in Paris, Grignon.

published in 2002. He addresses the issues of agriculture, food-supply, rural and environment policies, I have been working on over the last ten years. This is what I have retained, almost word for word from the work cited above.

—An agricultural policy is essentially a food-supply policy, for the great majority of agricultural products are used to feed people, and the quasi-totality of food products comes from agricultural products. The need to intervene in agriculture became evident in the United States, because of the natural instability of agricultural markets: the supply of agricultural products is extremely variable; whereas the demand for agricultural products is very stable, as they are essential goods. The quantity of agricultural products that is not consumed by countries producing them, and that is offered on world markets, represents only a few percentage points. It is made up of excess products, and market fluctuations give a false idea of world balances.

—The agricultural policies of the North are contested because they are very expensive, they generate production surpluses, declining prices and competitive distortions on international markets, and benefit only a handful of farmers defended by powerful lobbies. It is time to stop this waste, and to expose the agricultural sector to the laws of the market.

—The agricultural policies of all the member countries of the WTO, have been under international surveillance from its inception…but the global costs of agricultural subsidies of the OECD countries, has remained more or less the same as it was in the 1980s, a percentage of the subsidies for products has been replaced by subsidies for producers.

—Current negotiations tend to subject agriculture, food-supply and food safety only to the World Trade Organization which, by definition, has neither the competence nor the capacity, as it claims, to define the future of humanity's subsistence. As a matter of fact, the market is not the mechanism that is implicated.

If we take into consideration the institutional and political tendencies, but also the various natural, demographic, cultural and economic realities, what possible directions can be taken? Three scenarios can help give a better understanding of what I will now present from my own perspective.

—First scenario: the Liberal position carries the day. States intervene very little in the agricultural sector, and leave it up to the free working of markets to guarantee the safety of the food-supplies of their citizens, with respect both to quantity and quality. Those farmers who produce most cheaply in the world, will then conquer most of the markets: these are mainly the agricultures practiced in large estates of several thousands or tens of thousands of hectares, that use the most powerful farm machines as well as selected seeds, mineral and synthetic fertilizers, pesticides, and often employ farm hands paid less than 2 Euros a day. Such estates exist in many countries in South America, in South and East Africa, in the former Soviet Union, but also in Indonesia, in the Philippines... Under these conditions, a farm laborer earns less than 1,000 Euros a year, and produces more than the equivalent of 10,000 bushels of cereals a year. The cost of labor represents less than 1/10th of a Euro per bushel. The cost price of wheat is much less than 10 Euros a bushel, half of ours!

—Second scenario: international agricultural trade is progressively liberalized, but rich countries continue to subsidize their agricultures heavily by handing out internal subsidies, by guaranteeing minimum prices, and by reinforcing their policies of the safety of their food-supplies. They deliberately ignore the rules of the WTO, or cheat by using instruments that are compatible with the rules; such as remunerating farmers in the name of multi-functionality. Shrewdly managed, and combined with payment that takes into account the quality of products, subsidies can lead to forms of agriculture that respect the environment, national and regional development, and are more mindful of the food-supply needs that exist today. Following this logic, and overwhelmed by their agricultural budgets, rich countries decrease their aid, and simply let their agricultures die a slow death. As far as poor countries are concerned, where States have no means, hundreds of millions of peasants continue to be impoverished by the trend of dropping prices of agricultural products, due to the competition created by very competitive agricultures. The dependency of food-supplies that developing counties are experiencing, is getting worse; and poverty, destitution that especially afflicts rural populations, continues to spread, flaming civil wars and international tensions.

—Each of these two scenarios ensures the triumph of large rural estate agricultures with yield as the only consideration, using machines, seeds, fertilizers and pesticides, and employing low-cost labor. This victory is at the cost of the disappearance of agricultures, where nature or society prohibits such practices. This leads to the shrinking of cultivated, land while the world is not certain of being able to feed a planet of nine billion humans. As this cannot continue, one day we will undoubtedly reconstruct what has been destroyed. But at what price?

—Third scenario: The WTO encourages the establishment of real agricultural policies that are different from region to region; but encourage their autonomy, as well as qualitatively and quantitatively secure food-supplies. Alone, or organized into major regional markets, all the countries ensure protection against low priced imports at their borders, by possibly using certain controls employed by the former CAP. The level of protection of each market is fixed, so that internal agricultural prices enable farmers to live from their investments and their work. Protected from the competition that ruined them, and assured of a remunerative internal market, the peasantry of poor countries will begin to produce once more and, if we help them, to progress. Land and revenue taxes result in a certain type of equity. Import taxes, taxes on revenue and international aid make it possible to finance programs of agricultural modernization, and food vouchers for poor consumers. The rich countries set up policies that tend to produce less, but better, and turn towards the production of non-food products. To avoid perverse excess, product by product agreements with production quotas are signed on an international scale.

It is clear that I prefer the third scenario that ensures the safety and the quality of our food-supplies, favors the development of non-food production, and helps the most destitute countries protect themselves against international competition that is destroying their agricultures. This scenario enables those countries that do not know how to deal with their numerous peasants, to solve the problems they have. In France, it brings an end to this absurd situation whereby our safety depends on a budget with an upper limit that, nonetheless, because of burgeoning demands becomes subject to new expenses. This last point results in benefits to you, as consumers, in the form of compensation: agricultural products should be paid at prices that correspond to their costs. We will come back to this later.

—The international market is a market of available surpluses, where prices fluctuate in an unpredictable and aberrant way.

—Having run surpluses over a long period of time, and therefore maintaining low costs, this market could be in deficit with high costs for three structural reasons: the decrease in the cultivated areas of the world, climate variations that are increasing, and an increase in needs. Already appreciable today, these three conditions are evolving over time. It is reasonable to think that world prices will increase; forcing, but a little late, political authorities to intervene haphazardly, improvising all the while.

—By opening up the entire range of consumption to world markets with the current absurd prices, we discourage the development of production in developing countries; whereas they have an urgent need to ensure their global development, since in a quarter of a century the world will need all the agricultures of the planet.

—We open our markets to products that have been recently invented and have not been shown to be safe, as well as to products that come from everywhere, and do not necessarily correspond to our demands for quality. Forging ahead, the WTO forbids all qualitative criteria, declaring them contrary to free trade that remains its sole obsession.

—My last point concerns increases in prices that I deem necessary. Let me state five simple notions: 1) the relation of farmers to the economy, the perception they have of the reality of the markets, the feeling consumers have about the real costs of their daily living, the respect the elements that can satisfy our fundamental needs should get, everything is vitiated by a system of subsidies that destroys the link that must exist between the cost of production and the cost of consumption; 2) there are years when world prices will be lower than those that are the result of a policy that isolates our market, but there are moments when world prices will be higher than internal prices linked to our production costs; 3) the practice of variable prices that result from market fluctuations is what households have the most difficulty putting up with; they want reasonable and stable prices, our experience with the cost of automobile fuel demonstrates this fully; 4) paying European farmers' prices that correspond to the production costs of a good sized and well-run farm, would guarantee stability, and this is all the more so, as it would have only a slight effect on consumer spending. The price paid to farmers for their products represents only twenty to twenty-five percent of the retail costs, and the percentage of family food budgets has decreased to one eight of family income. Thus, an increase in farm prices can easily be made acceptable through reasoned policies. It remains that transitions must be put in place to prevent tensions that such a decision could cause, even if it happens to be absolutely necessary; 5) current studies on European Agricultural Policy, negotiations with the WTO, and the idea of a single subsidy per hectare, demonstrates very clear intentions: those in control are preparing for a worldwide trade system, and they will not tolerate any limits to liberalization that, in spite of all, threatens the food safety of many countries. This would lead inevitably to the destruction of the agricultural fabric and rural space that most of the countries in the world need.

—The return to the practice of prices, corresponding to costs, is the only step that will permit us to safeguard the safety of our food-supply and our rural space, our autonomy and our civilization. Hence, the choice you have as consumers is very clear: on the one hand, uncertain prices, quality that is not guaranteed, a territory in disarray and, on the other, slightly higher prices that are stable for the guaranteed quality of food-supplies in a cultivated space. As far as I am concerned, I certainly do not hesitate, and I suggest that you accept this as an objective necessity.

The policy I have just sketched is agricultural, it is related to food-supply, and it is environmental and rural. All of its multiple elements concerns, each and every one of you. It is much more likely to meet all your expectations if, when subjected to the vagaries of the climate, farmers are protected against the uncertainties of a market which, since it is a market of surpluses, fluctuates and is speculative. Farmers should be encouraged in the efforts they now are making in France to promote quality, protect nature and breathe life into four-fifths of our territory. This policy has the great ambition of meeting the needs of the living, both human and natural. It also meets the immediate needs of urban society.

Brussels

To the European Parliamentary

Agricultural Commission

A member of the European Parliamentary Commission that was hearing those who could provide some information on the eve of a major reform asked one of the individuals who played a role in establishing the Common Agricultural Policy to give his thoughts on his experience, as well as on the evolution that occurred from the time he had stopped dealing with theses issues. The following debate will permit me to go into greater detail, and I will content myself with rapidly enumerating a number of facts, propositions, then problems, of a political nature.

—The CAP must be rethought: 1) because having succeeded it has changed European agriculture; 2) because it has become an exporter, the EU must take into account changes that have occurred in the world as well as the demands of its competitors; 3) because it favors the richest regions and the largest farms through continuous subsidies; 4) because it subsidics products unevenly; 5) because in most regions the rural and farm populations have dropped below a critical level; 6) because the environment has become an important problem and a concern that the CAP has to take into account.

—The CAP must be redefined immediately, because of: 1) internal European financial commitments; 2) the dangerous evolution of international negotiations; 3) the fact that recently defined policies in the United States are very close to Europe's; 4) problems of food safety that are now coming to the fore; 5) the increase of membership in the EU that is complicating problems, and makes their solution incompatible with the budgetary cap that has been set; 6) the fact that many of our farmers draw more than fifty percent of their revenues from subsidies. We are used to this practice, but it is rationally untenable and is the object of international and intra-European criticism.

But before redefining it, we have to be aware of what an agricultural policy actually is.

—An agricultural policy is first and foremost a food-supply policy. Europe has become self-sufficient, only since the establishment of the Common Agricultural Policy. Demographic growth and our duty to feed all humans sufficiently lead us to believe that in twenty or twenty-five years the world will be unable to satisfy the population of the planet's food needs. The expectation of being able to meet the world's food needs must be qualified. It remains a constant worry since the techno-logical progress that is generally anticipated does not totally allay it. This should convince us that we need all the agricultures of the world to continue and develop, and our own in particular.

—By linking increased membership in the Union with a budgetary cap, the Union will have to reduce subsidies, even though there is no eco-nomic data that justifies this. By unlinking the two it would condemn agriculture to a state of increasing poverty that will be unbearable for many people. By unlinking the two, it would make subsidies to agri-culture seem artificial, unequal, and unbearable. This would amount to condemning the policy that has ensured the self-sufficiency of European agriculture that it had never experienced before, and to weakening its positions on the world market without in any way helping the countries that are truly under-developed.

—In order to attenuate the defects of aid limited to products, for a long time the European Union has opened its bank vaults, defined a regional policy and recognized the future of a multi-functional sys-tem. Together, these measures make up a complex but not always coherent ensemble. The time has come to establish a policy founded on principles of homogeneity that would also include modalities of differentiation that take into account the diversity of products, soils, and situations.

—Being at the same time agricultural, rural, environmental, and also dealing with food-supply, this new policy would ensure: 1) the pre-sent and future safety of our subsistence, and the quality of our food; 2) the protection of nature that has turned out to be fragile; 3) the preservation of rural life that accompanies and counterbalances the progressive urbanization of our society, that in fact safeguards our civ-ilization.

—The principles that guide international negotiations lead us to an impasse and to abrupt changes. Indeed, these negotiations tend to favor the countries whose agricultural production takes place mainly on immense farms that employ a quasi-servile work force. Since they favor the practice of low world market prices they overwhelm the agricultures of poor countries. But by following a model that has yet to be invented, they also oppress the rural poor and favor the urban-ization of countries and regions that need a long period of time to modernize. They force the United States to take convoluted measures, because the policies they administer are in contradiction with the principles they brandish. They destroy our agricultures and our civi-lization that retains the trace and the demands of the various soils from which it sprang. If the European Union would simply adhere to its own policies all of this would come about inescapably, today. Well, what's the use of choosing such a policy?

—But let's change perspective. A glance at the list of the members of your commission shows that, for the most part, they belong to agricultural organizations, or they represent agricultural regions. Your debates only deal with the problems related to farmers; whereas, today, it is necessary to link these problems to those of society in general. Agriculture is no longer a separate sector; it is part of an entity that finds its strength only in coherence; it must be perceived by all citizens as the guarantor of common collective interests. In the city, agricultural policy must no longer be perceived as a hodgepodge of privileges. It must be perceived as an objective necessity, imposed by the safety of our food-supply and by our concern for natural space. The agricultural world often annoys parliaments, mainly made up of city dwellers. Before opening public debate, your relations with the city must be worked out by taking into account the interests of "global society".

—We can reach one conclusion: To simply adopt, once again, our CAP and our practices is unsuitable, and even pathetic. We need a real reform. This reform must not be based on existing texts, but on a global vision of our needs; the needs of a continent that wants to be something more than a market; the needs of a world whose food production, and therefore agricultural production, must increase almost threefold in a quarter of a century, while everything leads us to believe the planet's environment is threatened.

—Can this be envisaged? It is difficult to do so for the following reasons: 1) the habits taken by our bureaucracy, and the rights acquired by certain professional organizations that take the initiative; 2) the contradictions that exist between you, whether they happen to be between your countries or your personal opinions; 3) the burden on the European Union's budget of a policy that we have a tendency to forget happens to be relatively heavy, because it is the only one in an ensemble that barely deals with social, educational, infrastructure, and safety problems; 4) currently, the Union probably does not have the political capacity to want and to negotiate a policy of its own with full-fledged States; 5) it refuses to consider its agriculture not as a burden but as an opportunity, and as safeguard.

Having presented my options in a text you already have in hand, I will be content to develop a few simple ideas. I will do so after having taken a few cautionary precautions: Obviously, the transitory measures that need to be put in place for new members, must be defined for a maximum period of ten years,. These measures must bring about a real unification of the market and a true harmony of the agricultural worlds that have taken divergent directions over a period of time. This must occur without useless tensions. Here are my options:

1) establish throughout the European Union a system of fixed prices in

relation to the real costs of production of good-sized and well-managed farms, but in conformity with the principles that respect family units of production;

2) put in place an internal and external policy of food aid;

3) for each type of product, fix quotas where the prices retained are respected; the export of surpluses would normally not receive any subsidies;

4) establish protection at the borders;

5) develop systems of insurances against fluctuations and natural disasters. These professional systems should be supported by public insurance mechanisms;

6) create a "European Commission for Approval of Food Products and Agricultural Practices". With the means to investigate and analyze along with the power to regulate the Commission can ensure the safety of food and of the environment;

7) define and support regional programs of "rural and environmental development"; these programs would propose remunerated activities to multi-functional farmers;

8) pay special attention to research and development of non-food agricultural products;

9) negotiate and conclude commercial aid and development agreements with African and Mediterranean countries;

10) just as what was done with respect to institutions, the constitution of a Council that would gather together the representatives of all the elements of our society whose mission would be to propose a European Agricultural, Food-Supply, Rural and Environmental Policy; a policy founded on stable principles that would be periodically and systematically evaluated and re-examined.

But the problem is fundamentally political.

—It is because it concerns our subsistence, our territory as well as the defense and the promotion of one of our activities. It is because it concerns the major balances of the world, and because the solution we bring to it will determine the European Union's place in the world.

—It has always and everywhere been a difficult problem to solve; and this is the case today more than ever before: because of the existence of enormous unsatisfied needs that are getting worse due to demographic increases; because of the fact that we are not certain that the world will be able to feed all its inhabitants; because of strategies of conquest that we have initiated and continue to initiate, even when they have destabilizing effects; because, increased membership in the Union of associated countries that have reached different degrees, even incompatible degrees of organization and development.

—It is political because of the level of subsidies that are given to certain producers, because of the budgetary costs these subsidies represent, because of the need to make consumers pay the real costs of their food, and we have gotten them out of the habit of doing so.

—It is because current negotiations put into question a certain conception of civilization, and force us to define the "European Model".

—It is all the more so, because it is a social and strategic instrument that is taken up by governments in total control, everywhere in the world; whereas in the European Union, it is administered by a complex system of shared powers. It is therefore because it urgently raises the question of knowing what Europe is; whereas the European Union has not yet defined either its ambitions, or its mode of governance.

The problem we must resolve is dramatic, and the European institutional system does not facilitate its solution. For the question arises of knowing whether the exercise of governance can make up for lack of government, or whether if it is unable to make major decisions a major human grouping can have a great destiny.

New York

To Representatives of the "Nations of the World"

From the point of view of an old novice, the international system seems very strange. Whereas the economy is foremost in the minds of people, it eludes the United Nations; it is monitored by the World Bank, the International Monetary Fund, and the World Trade Organization. Whereas safety is an extremely serious problem, it is monitored by the Security Council, and the General Assembly has only a consultative role to play. Whereas development controls the lives of billions of people and the future of world economic balances, it is the poor cousin of the UN; and the developing countries that are heard, are those that hope to compete with the prosperous Western World, not those countries whose populations are racked by hunger. Whereas goaded on by demographic growth, the best minds are attempting to find out if the world will be able to feed the world; it is an industrious but minor and impotent organization of the UN that is dedicated to examining this issue. Whereas climate disorders make us fear for the future of our planet, only stately High Masses deal with them, extremely well I should add, but with meager results because of the spectacular reluctances encountered. Any deviance from sacrosanct free trade is loudly condemned by the WTO; every monetary disorder justifies the intervention of the IMF; every tension can justify formal deliberations of the Security Council but hunger; yes hunger, climatic catastrophes, the danger for the world of a structural insufficiency of water, the destructive nature of waste that plagues the richest countries, the threat of the development of new endemic diseases result only in pious wishes whose respect is entrusted to the good will of each and every one.

I am ready to admit that an assembly of two hundred unmatched delegates does not have the capacity to deal with all the serious problems of the world. I admit that a fundamental reform is difficult if not impossible today, but I do not understand, and I cannot understand that a structural reform that would create a Council for Human, Economic and Ecological Development with deliberative power alongside the Security Council; cannot be adopted or proposed. The Council's mission would be to deal with all the systemic problems that need to be addressed for the future existence of the planet and the species, globally, and it would have the power to do this. It would be responsible for proposing a Code of Behavior to the Nations of the World, and if this code were respected it would ensure development not only of economic growth but also of peace, that is much more than non-war.

Having dared to express before you my ideas on the international organization and the needs of the world, I would like to develop the analyses that justify them, and the measures, above and beyond the reform of institutions, that could counteract the dangers that threaten us, along with the anguish that is beginning to haunt us. As I consider under-development to be an ill that, when it strikes a segment of the population, affects the vitality and tranquility of the entire world, I will end this demonstration by tackling the question of whether there could not exist; whether there do not exist, synergies between bad development and insecurity, especially today.

Why should I quote statistics that you already know? A single statistic, with the terrifying prospect that it projects on our inevitable future: taking demographic facts into consideration, and the fact that a billion people suffer from malnutrition today; to satisfy the needs of all human beings world production of food will have to triple over the next twenty-five or thirty years. It has doubled over the last half-century; can it improve its performance, while land is being turned into deserts, forests are being cleared, urbanization is spreading, increasing sea levels are threatening, are gnawing away at the world's cultivated spaces and that, if there are spaces to be improved, they need considerable investment? This gives rise to a large number of questions of which I will retain the following seven:

Is it really unthinkable and impossible to decrease demographic pressure that began as a promise, but has now become a threat? Can progress in science continue, without risks for the human species, and for nature? Can the Western World's model of production and consumption spread, or even continue to last? Can have-not countries today found their subsistence on imports; does this not increase their fragility? Do they have the will and the capacity to define and practice agricultural policies adapted to their realities? Do the modernization of agriculture and the continuation of rural poverty not both foster massive rural drift that cities cannot absorb? Do the development of exchanges and the market system not have the capacity or the vocation to satisfy all the needs of the world population? Along with real progress, is this development not already the cause of destruction and marginalization? But my last question: do these problems, that are intimately linked not force us to articulate a "political" vision of the world; and are we capable of doing so? After asking these questions, and before answering them, I would like to apologize for the blunt severity of these proposals. They have been dictated by my experience and by my awareness of my political responsibilities, and they are also possible because of the individual nature of my undertaking as well as my age that allows me to take these liberties. But let's get to the seven questions we asked, and first of all to the one related to demographics.

There is a large dial that is constantly swept by a needle in the entrance of a Canadian institution dedicated to international development. This mechanism actually counts the number of human births around the world: several of them per second. Eighty percent of these births take place south of a line that stretches from Gibraltar to the Dardanelles, then along the southern border of Russia and Siberia, crosses the Pacific with Japan, Australia, and New Zeeland to the North, and continues, following the border between the United States and Mexico. As we can see, demographic explosion is the daughter and twin of under-development. I had come to the Center for Research for International Development to discuss the problems arising from the decline in the African school and university systems. The poorer people are, the more they multiply; the more they multiply, the more they are unable to provide for their essential needs, including education; the less they are able to educate, the further they fall behind. Poverty, demographic increase, non-development, and under-schooling are the four faces of the specter that threatens our consciences. But should it be allowed, should it be possible for precautionary reasons that politics meddles with demography, when it raises serious ethical questions in each family, in each civilization and in each belief system? I will come back to this in a moment.

But as it is, and will be in the future, can the world feed the planet? Perhaps it can. But the world cannot be satisfied with a solution that would give the task of ensuring everyone's subsistence to a few already, or about to become, rich countries. Indeed, agricultural development is indispensable for under-developed countries that must devote their meager cash resources to acquiring machines, and not the rice, wheat, sugar or milk they need to live. But there is more to this: to feed everyone, the world today needs all the agricultures of the world. This brings me back to the second question: is it possible for the poorest countries to ensure their own self-sufficiency or to contribute to it? Without giving a long list of useless reasons, I would say that not all countries can satisfy the needs of their population through their own production, or by themselves, but all of them, and I mean all, can contribute to meeting their own needs. They must however do better, and do more than they are currently doing. But in addition to the economic reasons raised above, there are social ones: if an active state of agriculture does not exist, cities will be submerged by migratory currents that will subvert them economically, socially and politically.

Developing countries cannot deal with this rising tide because in today's world riches go to the rich. Since, to make them ever victorious over the poor countries, the rich ones subsidize their own agricultures. A reform of the international system would not suffice. Each poor country must work out a policy, and be helped contractually to establish it;

each poor country must be able to protect itself from the unfair competition I have just mentioned. Each country must have the means to develop its own research because the research the rich Northern countries have undertaken in the South has been done mainly for products that we ourselves consume. But the South must remain suspicious of the industrializing ideology that has driven its choices over a long period of time: it made these countries think that the land was a thing of the past and the factory the future. Agriculture and food production are the bases on which the Western economy was developed. They are the roots of the South's global development and the World's equilibrium.

Let's re-examine demographic growth, after this lapidary overview. It led to the doubling of the world's thousands of years old population in the last century, and this leads us to believe that the planet's population will increase by fifty percent over the next thirty years. This is the consequence of the progress made in medicine over the last half-century. But by saving the lives of newborn infants, the majority of whom died in the past, medicine has progressed much more quickly than social and conjugal behavior. Couples still want to have more children than they need to ensure they are looked after in their old age and to assure the continuity of the family. Figures show that these uncomplimentary rhythms will progressively balance out, but clearly the prolific generations and large families are the cause of the demographic explosion that we should be worrying about. Are there any means of intervention that we can use to contain it quickly? Certainly not! Can we intervene to end a life through abortion? The modern North believes we can, many churches oppose it and many couples refuse it instinctively. I believe it was a good thing that the North had the wisdom to regularize a practice that was spreading. I also understand that many refuse abortion, but I do not understand people who refuse to favor family planning that teaches methods of limiting the number of conceptions. I am incensed by positions that support a rising birth rate that gives life to hundreds of millions of infants, when the world today will not provide them with the means to live. No other assembly in the world can claim more legitimately the right to take up this issue; to demonstrate the diversity of the fundamental positions and lay out the elements of a philosophy that can reconcile facts with values. No country that is hostile to abortion can believe it is authorized to suspend all aid to the office of the United Nations, whose task it is to attend to the evolution of the population of the world.

No matter what demographic policy is followed, as it is currently forging ahead the world could experience a surplus population of two or three billion people in one third of a century. This is the problem facing the planet, today. It can be resolved, if the agricultures of the South are not subject to assaults from the North; it can, by favoring the development of research that encourages the development of all the productive factors of the South. Considerable efforts have to be made in these areas; for if the needs are enormous, the means are often pathetic, and the South is lagging further and further behind. It is lagging behind because the rich countries attract the best researchers, and these individuals who are often discouraged by the poverty of their laboratories, accept to go and find better working conditions elsewhere. It is falling behind, because research is a system and not a sum of specialists working in isolation, and in Africa in particular each country is too poor to enable research to reach the critical threshold when it can become promising.

It is falling behind, because it is in the interest of the laboratories in the North to offer the South "ready made discoveries" and that no one worries about the kind of research development policy that must be adopted. Starting from the analysis of needs that have been identified region by region, a research program must be put in place that would state the objectives, along with the human and material means; an international partnership program, that would put the experience they have acquired along with future perspectives in the hands of the South. Financed by the international community and assured of the disinterested support of researchers in the North, the absolute priority of this program would be to move ahead, starting only from the real needs of the regions that would serve as a framework. As ambitious as School or University programs; such a program is indispensable to facilitate the creation of an autonomous and responsible South. But the North must be convinced that this is also in its own interest. Will it one day be coherent enough to admit that the real development of the South doesn't threaten it, but on the contrary, that it will create thousands of opportunities? No matter how it turns out, the destiny of the poor controls the future of everyone. At least this is what I believe.

But things become complicated when we try to make headway in defining development. The production and consumption model of the rich seduces the world, and it does so because of its remarkable results in production. It does so by wasting the limited resources of the planet, and no one seems concerned about this evidence: the resources on hand and those promised by science, do not permit nine billion people to consume as much water, pure air, energy, food, as are consumed by a billion Westerners today. We can count on progress to "loosen the belt", but no one can believe that all the countries in the world will be able to reach a GNP per inhabitant that in any way approaches the one the average citizen of the fifteen countries of the European Union currently enjoys. To call for these countries to develop without stating this amounts to lying to Westerners and to the others; to the rich, and to those who are not. We must verify the hypothesis which, after all the estimates, indicates that unless new models are invented, and even sup-posing they are invented, the equitable satisfaction of everyone's needs can be met only at the expense of the lifestyle of the rich and by limit-ing the hopes of the poorest people on earth.

The international system can tackle the problem and attempt to resolve it, if all the agencies work together. We could then elucidate the major agricultural uncertainty that handicaps the future of the world, and we could cost the material, intellectual and organizational invest-ments that would enable us to overcome it. This is a huge ambition, and so is the task facing us. The world can feed itself only if the following policies are put in place: the modification of the policies of Europe and the United States; the right of poor countries to protect themselves at their borders; food aid in return for investments in development; aid to establish local and national markets; aid to cottage industry; to the orga-nization of producers; to training; then to research; and putting in place the necessary public infrastructures; the equipment of villages.

This would be a titanic undertaking, if we were not certain that the peasants themselves would reap the fruits of their efforts and they would also begin to produce once again. They would, for if we take only the example of Africa, yesterday the continent was self-sufficient and today it aspires to become so once again. It can attain this goal. Once the human spirit is rekindled everything will begin to come together again. To rekindle it every peasant must know that his family will benefit from his effort. Regions are not poor because they cannot exploit their resources and create riches, but because conditions are such that they gain no benefit from their resources.

But what is the use of hiding the fact that, no matter how great an effort is made to encourage agricultural development and the peas-antry, a daunting problem still remains that no one knows how to resolve. To understand its signification and its scope we must remem-ber that the development of the most efficient agricultures of the world was accompanied by rural drift that also enables and favors industrial development. Today, factories no longer need great numbers of labor-ers, and the development of agriculture does not mean that large num-bers of farms have to disappear. We have to ask what China, India, Africa, Latin America, and Indonesia will do with the hundreds of mil-lions of peasants who will not find a place in either factories or cities, after being chassed from the land by the modernization of agriculture. What can we do? It is time to think about this, and to work to find a solution. The problem is of significant importance and is fascinating at the same time. We must invent a civilization that melds the advantages of the city with those of the country, and combines the synergies of a modern farming agriculture with a society that develops in a rural envi-ronment. We need nothing less than this.

This is the second to the last question, I want to raise with you. A final one remains that I will tackle without any useless precautions, by highlighting the risks the world runs along with the necessary political stakes it entails; by highlighting afterwards, why you cannot hide behind your statutes as the reason for not examining this problem closely.

Considering the unsatisfied needs and the increase in population; in the next thirty years the world will have to produce three times more food than it currently does. Whether or not it succeeds in curbing the demographic explosion, and even supposing it wants to, it is not certain the world will be able to do so. This can depend on technologies and science, but it is also needs to take adequate precautions to control them, even so it is not certain that science and technology will suffice. They can suffice only on two inescapable conditions: the Western model of production and consumption must be tamed, because it is wasteful all the lands and all the agricultures of the world must be prepared for production, because the poor world cannot afford the luxury of importing its subsistence when it needs equipment.

Because you are the forum of the world, and because you are the meeting place of all the continents, you should be the epicenter of these interrogations, these studies, of the choices that can direct the necessary policies. Who will do it if you do not, and what will happen if you rely on world currents and mankind's good will? The real world is not pursuing the trajectory it was familiar with when the United Nations was founded. Information circulates; images multiply infinitely. Messages are exchanged that create communities of ideas above and beyond borders, and give rise to hopes. These messages have a tendency to transform into hope what was fatalism and resignation, in the past. The time has come for the General Assembly of the Nations of the World to situate the problems of development of the territories and of human fulfillment, at the very core of its objectives. It is through shared and equal responsibility and true collaboration between the Development Council and the Security Council that world peace can be assured.

We must fear a world, where the toll of security would ring the death knell of those left behind. Let's try and understand the vibrations caused in the Third-World by the battering that shakes institutional and financial citadels. We must hear the calls to revolt that mobilize the hopes of the poor. Not having the audacity to promise an equal sharing of the riches of the world, we must try and replace the binomial, waste-destitution, today, by the binomial, comfort-poverty. We must accept inequalities, as long as they do not threaten either life or human dignity.

Versailles

To Researchers

I am not a researcher and I am unaware of the essential progress made in science, yet as a citizen and a human being I need to say in front of you what I think about everything you have done, but also about what you are currently doing. Caught up between wonder and fear, between hope and dismissal, I feel it is necessary to inform you, scientists, about the dramatic hesitations that an old man fascinated by progress experiences when faced with what Alain Ruellan designated as the main stakes of the next century: the gap that separates the power of the human race and the wisdom it can muster when exercising this power.

But before dealing with this issue, I would like to offer up a few brief remarks. The first is related to the risk of intellectual isolation from others that may be inherent in your profession. Without going so far as to say that a researcher is a person who knows more and more, in a narrower and narrower area of specialization, I would like to invite you to practice multi-disciplinarity, not only in teams working on projects, not only in commissions, but in life itself. I was told that in the dining halls where you have lunch, each table is made up of researchers in the same discipline, who continue to work when conversing. Abandon your own areas of expertise, discuss, chat, read outside of your areas of specialization, and you will understand better. You will communicate better with public opinion that your science fascinates and frightens, at one and the same time.

Let's examine some areas of research that I tend to make a priority, in politics. I will begin with a few examples: Knowing that certain human organisms are allergic to lactose, you have succeeded in developing milk products that without being entirely free of it, have low lactose content. After you took into account the ills represented by obesity you succeeded in identifying the foods and the abuses that cause it. We are therefore led to believe that using the progress made in genetics, it will soon be possible to give all individuals a card that lets them know what suits them, or what is inadvisable. This would be real progress for everyone and, at the same time, it would save the social security systems an enormous amount of money, dietetics would become preventive medicine! The stakes are such that this merits a real effort.

Many of you undertake your specialized research from the perspective of hunger that exists and that is spreading in the world. Progress here has been notable, and research to overcome hunger is still going on. But all of your discoveries will not make it possible to understand and bring an end to hunger, especially in the rural regions of have-not countries. In infinite guises, hunger appears as a state but also as a dynamics where natural, social, religious, cultural, economic, administrative, technical and political factors intervene. For hunger to disappear, it is not sufficient to develop nature. The time has come to develop a typology of hunger, and a dynamics to deal with it. The time has come for them to become the object of research between disciplines that normally do not communicate with each other and to elaborate scenarios and means for dealing with hunger that are better adapted to real local situations. We will never overcome hunger by sending aid, even seeds and experts, but we will do so only if we can mobilize the factors that taken together will give the starving masses hope that through action they can escape what they consider a fatality. By mobilizing those who are hungry against what they consider as inevitable today, and by inventing complex dynamics that will enable entire territories to overcome their state of resignation. Political will is no more likely to bring this about than aid is.

Moving on from areas related to feeding the starving, and thinking about the debate about the competition of our subsidized agriculture with the agricultures of developing countries, and thinking about the exhaustion of fossil energies that has been predicted, we must favor the production of plants that can provide fuels but also bio-degradable fibers, as well as other non-food products (cosmetics, drugs, etc.). And, I imagine that this has already been done: we must encourage the methodical inventory of research that can help us deal with food, or the other shortages that threaten us.

Going on to another area of concern, I would like to mention here the evolutions and ills that rural societies go through and experience, and to tell you how pleased I am that the social sciences now play a major role at the National Institute for Agronomical Research. So much remains to be done. I do not want to get carried away in front of you by the worries, the disarray of farmers, in particular peasants, causes me, yet, I suggest that together researchers address the problems raised by agricultural policies, such as the one opposing productivist agriculture, that is mainly interested in performance, and biological agriculture, interested in health and environment, from the perspective of all the disciplines. The time for disagreement is over; it is now time to identify those efficient agricultural practices that respect the natural environment, and the concerns consumers have for quality. This is an old idea that was championed by Jacques Poly, formerly a strong advocate of an efficient and autonomous agriculture; Guy Paillotin recently gave a report that was both interesting and useful, on well-thought out agriculture. This is wonderful, but if we take into account the assortment of scenarios mapping out possible futures presented by Philippe Lacombe from the perspective of agricultural macro-economy, it is urgent for your research teams to identify the productive systems that will help us escape the frenzy of quantitative measures, without abandoning productivity. The decision-making political leaders must have rules that base the subsidies they hand out to farmers on facts; in this way we could avoid privileging those who need it the least. These are difficult but necessary initiatives, because the world will soon have to feed nine billion humans, and it will be impossible to resist the most reckless innovations, if they appear to be the only ones that make it possible to feed the world.

You have studies that clearly demonstrate we do not have to link economic performance to animal or vegetal forcing. These studies show that in Brittany, Normandy, or in the Vexin region of France, it is possible to produce enough, cheaply enough, without endangering rural society, nature and public health. It is time for this to become an area of research, with clearly defined policies. This is urgent, because the agricultural policies of the European Union and the United States of America hand out subsidies without any concern for real needs. A policy with global objectives, through the market, or through intervention, would give people what they reasonably need to live well and carry out all their tasks. I am not thinking about a simple mapping, but about establishing a typology that would enable those distributing aid to know who needs what to carry out his or her various duties. We are very far from this state of affairs. To prove this, all we have to do is to examine closely the relationship between the subsidies distributed and the work done by those who receive them. The people who decide must know what interventions correspond most to the needs of society and nature. I speak from experience, having made mistakes about this in the past.

But the list of the research you should pursue and the research you should undertake, remains long. I will focus on the domains where researchers in the farm-produce industry are vigorously concentrating their efforts. It is imperative for laboratories to determine what is dangerous, what is doubtful and what is good that should be created. This is probably where public research is most indispensable: where it can probably better judge innovations, society needs to be informed. In the most advanced disciplines, all progress made raises problems. Ethical committees have confirmed this. In spite of their great merit, they cannot address all the problems. They must be able to prevent certain kinds of research and to do this, public research has to march in tandem with private research, and they also must be able to say where the research being undertaken is leading us. In this way we can resolve issues related to the means and the status of public researchers. They must not either be tempted by, or have permission to, or the occasion to become the accomplices of colleagues who consider every profitable innovation to be a good thing.

On a European scale, one could therefore create a commission to evaluate agricultural practices and products. This commission that would have the power of investigation, the capacity of analysis and evaluation, could warn, forbid, and even take to court. Your institutes would be its privileged councilors just as in the USA public universities and laboratories are the councilors of the Food and Drug Administration. This would be an immense progress over what we have, for a recent study showed that public, or university laboratories prove the harmfulness of products they are asked to examine four times more often than do private laboratories. Such a mechanism would have the dual advantage of assuaging the fears that progress inspires, and of giving democratic debate and political deliberation the legitimacy and the reliability they need. We have reached the point where it is time to raise ethical issues.

In your personal thinking and your work, in your laboratories, in your research teams and meetings, do not think only about the secrets you can pierce and the progress that can ensue; also think about the harm that their application can bring about along with the terror they can inspire. Having abolished the sacred, ask yourselves what society can accept; and tell us how, together, you could define an ethics, and how you could take an oath that would protect us from the fear of progress. You should dread the revolt of the Canuts[1]: if it did happen, it is not be only the looms that would be destroyed but the alliance that I speak about from experience that over the past two centuries has linked Western society and civilization, on the one hand, and Progress, on the other.

The contradiction arising from the progress that you have initiated, and the Progress that all society will benefit from, reveals a deep malaise. One day, will you not collectively, or individually, experience the moral drama that Oppenheimer, the co-inventor of the atom bomb, did? Don't exclude this possibility. Genetics and transgenesis, the GMOs, certain fertilizers that pollute hundreds of thousands of hectares can provoke disasters comparable to those of Chernobyl. Because science has brought us, and continues to bring us the best, ask yourselves how we can avoid the risk of it bringing us the worse. You know that no matter how harmful it happens to be a discovery that can result in a profit will be exploited, you also know that the law can prohibit research only when it is too late and it cannot prohibit research from being undertaken. Help us wisely to understand, evaluate, accept or reject what you cannot control by playing the role of a sorcerer's apprentice. Always make certain the excitement that can result from

---

[1] The Canuts were silk-workers in Lyon France who revolted at the beginning of the XIXth century in order to better their living conditions.

finally piercing one of nature's secrets is accompanied by the profound joy of scientific explanation that will help find a solution to a human problem.

Perhaps you will be surprised by my preoccupations that occur from progress being made today. Let me give you a few facts spread out over a forty-year span that make me suspicious of the ambition of researchers whose passion for knowledge is motivated by the lure of profit. Forty years ago, I visited some remarkable laboratories in Switzerland. A slogan printed on the walls greeted our delegation. I will quote from memory: "Nature is the first producer of vitamins in the world. Roche is the second". Well, I said to myself, nothing seems able to stop them. When I was recently in North America, I heard a woman scientist announce the birth of a human clone. I instantly detested this. The next day, some scientists expressed doubts about the ability of this young being to survive. Two days later, the same scientist, all smiles, announced the birth of a second clone that was identical to the first; they died a short time later. Not too long ago, under the title "After terminator, here's Cancerinator", the *Courrier International* published an article from the *Observer* about some researchers who imagined introducing rapidly mutating genes in plants and animals to study the accelerated pace of their evolution. And, too bad about cancers that will inevitably result from this. What is frightening in all of this, and with the GMOs too is the speed with which firms want to launch into the world what their research teams have developed at great costs. I am not in principle against the GMOs, but we need time to let us know what they represent in terms of progress and risk. Are we so vain as to believe that our failures will yield so much information that we will one day attain perfection? Thinking about the wisdom of the Greeks, I tell myself that by forcing the secrets of nature, we demonstrate our pride; we step beyond the bounds, and run the risk of unbridling Nemesis' vengeance.

As we are progressively losing our respect for the sacred, who will explain the road to wisdom when we are tempted by the fury of science? Here again, let's take the following example: because she received generous help, a young Mexican girl benefited from two kidney transplants; after the first one failed the surgeons began anew. They even went further, and a medical bulletin announced the success of a third transplant, also informing the public that the young Mexican girl was encountering serious cervical problems. In the meantime, she died without anyone telling us about her dreams and her suffering, or her agony. Intellectually, I can admire all the efforts that were made, I do not doubt for a moment that one day surgeons will succeed where they have failed today. I am certain that the team of surgeons wept for this young girl, as it would for their own children. Nonetheless, I cannot ethically, or even intellectually, condone this surgical determination to continue operating. More generally, I cannot, without precaution, give my human support to what I am tempted to consider as a form of relentless scientific activity.

To make myself clearly understood, I would like to sketch a plan by underscoring the logic that could serve to organize a Museum of Natural History. It would be divided into five sections, beginning with origins and ending with the present and predictable times. The sections would be labeled in turn: "nature before humans", "humans in nature", "humans and nature", "humans against nature", and finally, "for the rebirth of nature". The first and second sections would together cover a period some millions of years long, and would present nature at its beginnings; then the appearance of the human species living among the other species, feeding themselves by fishing, hunting and gathering; like the other species. The third section would span a period of tens of thousands of years, and would deal with tribes as they became sedentary, with culture, with the raising of animals, then with irrigation, selected natural selection, grafting, artificial insemination; all of this occurring in a natural environment that has been developed, on soils enriched by natural fertilizers. In short, the world as it existed fifty years ago.

Spanning a few decades, the fourth section would be the one we are currently experiencing. Disrupting what has been acquired, not taking the time to think things through, human activity both responds to the demographic explosion and favors it at the same time; driven by competition that favors performance, motivated by public strategies that are seeking power, crossing scientific discoveries in agriculture with those of chemistry, of medicine and genetics, manipulating nature even to the point of human and non-human transgenetics. When human activity throws out a challenge and seems to refuse any limits whatsoever, it creates a problem that it cannot solve by itself. The fifth section would be devoted to reconciling the curiosities and needs of humans reconciled with nature. In one way or another, in this museum, scientists and museologists would figure out what acquired, or about to be acquired, knowledge would permit society to realize, both what is good and what is bad. The originality of this trajectory would be to tell the story of our history, and not to present objects or periods in time. Highlighting natural or willed logics and dynamics, using wonder and fear, it would be organized to teach children, adolescents and adults in such a way that, once aware of these successive evolutions and after becoming conscious of progress in the world, citizens would make choices and act in a fully informed fashion.

In a few words, here are the problems facing our generations:
—Progress, especially scientific progress, has resolved problems of subsistence and health, and those who have access to them live longer

and better.

—The efforts of researchers that for a long time were devoted to a better exploitation of natural riches, and even improving on them, are now focusing on bypassing nature through cloning, and especially transgenetics. Having pierced the secrets of nature, they want to reshape it.

—The progress made over the years is accompanied by promises and threats that are true, or exaggerated by public opinion, and for the sake of rationality it is protected by the market.,

—Certain types of research and discoveries challenge essential ethical values that are no longer guaranteed by religion for everyone.

—Hence the question arises of the need to ban and, in this case, of the need to designate those who are empowered to say what is forbidden and to also say what sanctions can be applied to those who transgress.

—Through its legislative function, the Political can define the elements governing process but as it is unable to predict it, it cannot define case by case the substance of what is forbidden. It will happen in the future, and it is unpredictable by this very fact.

—But the Institutional Politic cannot act alone. The choices made can have consequences on civilization and on life itself; and only a fundamentally democratic process can give the necessary legitimacy to the directions that have been retained.

—The scientific community is also bound by this, for only it can predict the consequences of the implementation of the advances that have been made, or are about to be made. As this community is undoubtedly split on such issues, it must clearly state the positions that have been put forth by its members.

—Ethical Committees are an undeniable step in the right direction. But beyond this, we must put in place a system that links scientific deliberation and democratic process with legislative and juridical definitions.

—Enlightened by the scientific community, society alone has the legitimacy and the power to ban.

But the world is evolving:

—On September 11 2002, the European Communities' Court of First Instance affirmed the primacy of the principal of safety under the advisement of experts. It brings into play political responsibility, but on condition that the decisions made have a democratic legitimacy. To take a decision authorities do not need a complete risk analysis, they must demonstrate "in an informed way" the "probability" of the risk they refuse to assume. This covers risks linked to the hasty implementation of products that can represent dangers. The decision of September 11, 2002 establishes the principle that society, the body politic, controls the dynamics of science.

—Without saying so expressly, the recent decision tends to make society responsible for the living. This decision invests society with the power to say what must be protected. It gives democratic process the responsibility of defining what, even in the recent past, belonged to the domain of the sacred. There has never been in history a greater consecration of history. We have to found the legitimacy of a democracy that is now responsible for defining life. Having liberated itself from ancient wisdom, science now forces us to discover modern wisdom.

Ladies and Gentlemen,

Your community of scientists is now Society's only councilor that can define life, and democratically make the choices that will determine our collective destiny. Individually and collectively, your moral responsibility is immense. It must be a source of obsession without creating a sense of paralysis.

Clermont Ferrand

To Young Farmers

Although I am not a farmer, nonetheless I am as attentive to agriculture as you are, and my passion for agriculture is as great as yours. I ask as many questions about it as you do. I follow agriculture from the distance of politics, history and philosophy, whereas with your boots planted in the earth, driving your tractors, in the middle of your herds, your eyes following weather reports, and your accounts, you now feel condemned by a society engaged in a frantic race. I could oppose your views, but what would be the use of doing so? We could debate our views, so that a project would emerge where the needs of society at large and yours would both be linked and complement each other. This is the only way you will be able to have a life worth living, offer a real future to your children, ensure the subsistence of everyone, and become the artisans of a civilization that respects nature. But to get there, we have to change logic and politics.

Forty years ago, I came and told your elders that we needed to produce more and that all of our efforts, theirs and those of the Nation, had to contribute to increase production. The politics of the structures itself that they claimed anchored their very roots would favor an increase in production. This objective was easy to define because it was dictated by circumstance. We worked together and we met it. But we did not notice the incompatibilities that could exist between the objective of parity, dictated by the Laws Governing Higher Education, and the objectives of European policies that were governed by a global economic strategy. We especially did not imagine how much the world was about to change. In discussing our errors or, rather our lack of foresight, I can say that in wanting to produce more we did not get our objectives wrong, but we badly evaluated the consequences of this aspiration. Contrary to what we imagined, to produce more made us take a direction that led to the future disappearance of two-thirds of the farms and of more than two-thirds of the agricultural population of France. But what would we have done; what would I have done if we had been aware of these consequences?

I never believed that this increase in production would happen with the same stable population. The creation of the Social Action Funds for the Development of Agricultural Facilities (SAFDAF) confirms this. But while even accepting a certain amount of rural population drift, I never dreamed it would reach the proportions it did. Had I been able to imagine this, I would have been tempted less to oppose it than to correct its human, social and territorial consequences. But

159

what's the use of complaining about this. We should take note of what has happened, and now try and work out a policy that stems the current population drain and founds an agriculture without subsidies that thrives on its own production and the multiple tasks for which it is now responsible.

Having refused a regulated economy, we should reject a system that does not allow a good-sized and well administered farm to balance its books only from the sale of the foodstuffs it produces. We must put market ideologies under close scrutiny and refuse to be mesmerized simply by production. We should be aware that today agriculture has three functions: it nourishes us; it safeguards our natural environment; it links and provides living conditions for over four-fifths of our territory. In the multiple roles it has, it fulfills many functions: by its economic function, it belongs to the primary sector as a producer of raw materials; and it belongs to the manufacturing sector, because it often transforms products to make them immediately consumable; but also to the tertiary sector as the purveyor of various services, some of which are market driven and others are directly related to public service. These functions are a burden, but also a privilege, because agriculture is no longer a separate world whose idiosyncrasies can be supported by politics. Agriculture is not an activity that is only market driven. Your responsibilities and constraints must be articulated, along with the agreements that need to be reached for you to be able to deal with the issues you face.

But the nourishing function of agriculture is its primary activity; this function has a great future, because to feed the nine billion people who will be on earth in a quarter of a century adequately, the world will have to produce three times more food than it does today. It will have to, but the transformations needed to bring this about will be difficult to initiate. We must produce as much as we can, while respecting the environment; and our production must not prevent the development of have-not countries. Our agriculture must develop within the framework of interregional strategies, and must prepare for a future where there will be technological progress, along with better and more successful organization and strategies. Our agriculture must devote a part of its investments and its activity to the production of "fuel", and other non-food products the world needs. There are, and we have to be aware of this, needs that must be dealt with urgently. But these needs vary; and we must organize ourselves to address them, at home, in Europe, and in the world.

But in addition to their function as producers, farmers have other tasks they can carry out better than other sectors of the work force; and society really needs these to happen, either through necessity or inclination. You have a number of tasks to carry out: putting an end to the pollution caused by your farming methods; instituting proper water management; ensuring the upkeep of the rural landscape considered not as simple images but as living realities. The problems of pollution can now be dealt with because of the evolution of technology. By solving these problems everywhere, our worries about nature will not be confined to a few protected zones. We must look after our entire territory. When certain practices are forbidden, we must make those who flaunt them pay the price. But, above and beyond pollution, the upkeep of such a space requires work, and those who do it must be rewarded. You must devote a part of your labor and of your time to these activities.

But there are other tasks, more of a craft-industry type and linked to agriculture, that open up new horizons that are greater and greater because the food-supply market is diversifying. There are networks that have been built up within the worldwide logic of productivity, but these concern basic products. There are areas where quality and specificity carry the day. These areas of production have important prospects and are not unimportant from an economic perspective. Producers must organize themselves in such a way that their concerns for originality do not isolate them from each other. These organizations must help them find buyers, not only in regional markets, but in Paris and around the world. I dream about being able to buy on the Internet the wine, cheese, jam, dry sausage that I discovered over the years in the markets of France in Sarlat, Uzès, Saumur or Cluny. Only an organized group of farmers can make this happen.

There are other tasks of public interest that offer jobs of various types; albeit small jobs but innumerable and available to everyone. We must return to the times that inspired the inventors of "territorial development contracts" and make these the tools of development rather than instruments to allocate agricultural capital. There is tourism and other activities that may or not be linked to the farm. This is how more than half the national territory lives with difficulty, but it also lives well. It is worthwhile for you to concentrate on this; it is worthwhile to pour more help into this area, because the future is uncertain. There are, finally, outside of agriculture all the activities that would like to be transferred to rural society, because their managers suffer from the ills of the city. But in order to decentralize firms, especially small ones, they need a social life and conveniences that cannot be found everywhere today. Several areas have to be taken into account: basic agricultural production; typical products; their transformation; the environment; the territory, and social life. Tackled all together by you, at home and immediately, they would permit you to invent your own modernity rather than be subjected to the modernity that threatens you.

Indeed, the future is hesitating between pursuing blind and deaf globalization and taking into account the diversity of the living.

Those who are calling the shots today have a very simple theory: the more and more strictly their laws are respected, the nicer and better the market will be. I reject these notions, just as Joseph E. Stiglitz, Nobel Prize in Economics does: "Today, globalization does not work. It doesn't work for the poor of the world. It doesn't work for the environment. It doesn't work for the stability of the world economy." It doesn't work either for you, or for the Farmers of the Mid-West. It works even less for hundreds of millions of peasant families in China, India, Oceania, and Africa. It doesn't work, and they want it to continue. Now, if this were to continue one or two billion peasants would migrate over the next few decades to cities that cannot accommodate them. It doesn't work and, except for a radical change, things will go from bad to worse. Change will not occur by itself and it will not give rise to the implementation of a standard model. France is a proof of this, and the world is much more diverse than this country is. It is not sufficient to be against standardization, we have to develop ideas to counter it.

But even if you are not alone, you are not united, and you don't have a global vision to oppose market ideology. You are the enemies of those who share your fate because, the market being what it is, you have to overwhelm your colleagues in order to take their place. Peasants – artisans from the Massif Central and farmers on the North Plains, you are fighting against one and other, not on the markets because you do not produce the same thing, but in the halls of the Minister of Agriculture, or of the European Commission, to share subsidies that are getting more and more scarce. You want to protect yourselves against the Maghreb, the Americas; in truth against all the continents, for almost anyone can produce anything with the help of technology. If market dynamics are so certain of the future it is because they question our capacity to develop a concept, an organization; a vision that is more pertinent and more globally economically efficient.

Hence, the future is indecisive. But what it is indecisive about? On the one hand, there exist a doctrine and a mechanism, money and power. On the other, human truth, a refusal of good arguments, ideas, but none which map out another future. The market will continue, because sharp criticism and unbearable facts will not make it put itself in question. We have to oppose a well crafted image of a world that works for people and for nature, to the ideal image it creates of the laissez-faire world. This is all the more difficult, because this image must oppose the complex to the simple. It is simple to say supply and demand, competition, free trade; since these different words, along with many others, only spell out a single law, a single regulation, that of the market. The mathematics of accounting that sometimes go wrong, however; are thought to be sufficient to ensure the happiness of the world. Drinking water itself becomes a market commodity for only those who can pay for it; too bad for the others. Our vision is more complex, it is as complex as life itself. And because our vision is complex, it is difficult to explain, and therefore to have others share in it, even if the great majority of human beings instinctively adhere to it. We are aware of the negative slogans, but our vision has not yet come up with some slogans that are both positive and relevant. I shall try and state in positive terms the complex vision I have of the world.

All members that make up human society must take part in this reflection, because everyone is concerned. This is a difficult enterprise, because of the large number of diverging interests and points of view; but it is possible to succeed, because of the deeply-felt anguish caused by the actual evolution of things. This undertaking is necessary, because it is the only one that can put an end to a market monologue addressed to a clientele that has been elevated to the status of an actual society. This clientele is often unsatisfied but passive, because it is a prisoner, for want of an alternative. What can be done about this? Let's begin by identifying the difficulties; then suggest some avenues of reflection and action.

First of all the facts: globalization does not work, neither does protectionism; the market ignores or tramples what it does not favor, a regulated economy does the same thing; your government defends you badly; you, in turn, do not defend yourselves except by demonstrating against the government and asking for subsidies; Europe disappoints you, you say you cannot do without it; administrations have no ideas, most of your professional organizations don't have any ideas that are worth anything; even when they disappoint you, you renew to the bitter end your professional managers; even when you find the future uncertain, you abandon yourselves to it; the progress of science frightens you, but the majority of you take advantage of it indiscriminately. You grumble and you accept without waiting for the great day, because you know that it will never arrive. You struggle like people possessed, on your lands, in your stables, in your assemblies and, when you get personal results that are sometimes surprising, you end up by resigning yourselves to the situation, thereby joining up with the fatalists of the Third World.

The first suggestion that I would like to put forth, deals with method. Taking into account the quasi immobilization that seems to characterize institutional systems, I believe the answer can be found only in society, or its components. However, I would like to suggest you don't work alone but, instead, that you compare your needs and your ideas with those of others. It would be necessary, it is necessary that, everywhere, small inter-professional and civic teams come together and reflect on the needs they feel; on the fears that each of their members experience; on the contradictions that need to be resolved; on the experts, the wise persons they could consult; on the relations they could establish thanks to the Internet; on the concrete and clear texts with questions and proposals it would be possible to compose; on the contacts, based on all of the above, that it would be possible to make with Québécois, Walloon, African, Caledonian, Swiss, Martinique populations to exchange ideas and reflect together on the analyses and proposals, but not on resentments, revolts and protests. In doing this you will learn that you are not alone in having to deal with these problems; and you will help others to discover and understand you.

Virtual seminars at the levels of the district, the region, the country, Europe and the world, would enable us to ask the right questions and suggest new answers. Suppose, that instead of a lot of foolishness, the WEB enabled us to circulate the elements of concrete, free and multiple research, without prejudices nor leaders; and suppose this works, do you not believe that after having tried to take it over, the economic, professional, administrative and political leaders would end up by listening to the immense, calm, demanding and nonetheless promising rumbling. And if "this didn't work", you would have to begin afresh, tirelessly, because that's the solution.

If you took this route, new perspectives would open up in every direction, in every sector, and also in the political domain. There would be a new form of democracy everywhere; not the democracy of power that institutions found and organize, but the democracy of debate and social inventions that the institutions don't know how to sustain. Why take such a long detour? Because no agriculture can get there on its own. Because, even together, none of the agricultures will get there, if society as a whole doesn't understand that its future depends on them. Because France, Europe, the North, will not find peace alone, unless the entire world takes this on.

Let me dream that, tonight, before leaving, six friends present here who are from the six corners of France, will exchange email addresses, and each one will decide to begin a discussion with one or two neighbors, one or two city dwellers who live near them; not on capitalism, socialism or globalization; nor on democracy, nor on Brussels and the subsidies handed out, but on their actual daily needs. Then, slowly, patiently; like old peasants, they extend their topic of discussion; taking into account all the areas of France, they exchange questions and, if it happens, the discoveries made with their friends here tonight. Slowly, and irresistibly, the face of the world would be changed by this. It would be, because, step by step, you would have become the actors of your own destiny in a world that seems subject to the power of anonymous dynamics that can be controlled, only by exercising true democracy.

Let's turn now to content, or what I called vision a few minutes ago. Let me now outline what was worked out, as I read, listened, analyzed the numbers, conversed, discussed the essential elements. This will be my political answer to the Report on the Laws Governing Education, adopted in June 2001 during the 35th Congress of the National Committee of Young Farmers (NCYF): "Peasants of the World: The Price of Our Future". I studied this report with a great deal of attention and, as you will note, I often agree with it.

—The need for food-supplies in the world in a quarter or a third of a century will be three times more important than it is today. To be able to feed everyone on the planet will take a worldwide concerted effort. The future of agriculture is not threatened by inadequate needs, but by a lack of vision and by the blind workings of market dynamics left unto itself. Only a true policy can ensure the respect for everyone's legitimate expectations.

—The future of agriculture lies in its capacity to ensure our survival. It also lies in bringing solutions to the serious problems caused by its very evolution. It cannot solve these alone: for these problems are often of common interest, and to solve them agriculture must join forces with those who are affected.

—There is no longer any room for an agricultural policy negotiated between only professional organizations and institutions, whether they happen to be national or European. We need a rural, food pro- ducing, environmental agricultural policy that will reestablish the essential link between the agricultural world and society at large; that is concerned with production, but also with everything that makes it happen, and in which it happens.

—It is no longer possible for the most efficient farmers to earn half of their revenue from subsidies that compensate for the lack of profit resulting from worldwide price levels. We must therefore separate the internal European market from the world market; set internal prices that take into account the real production costs; these prices would be guaranteed only for the quantities necessary to meet the needs of the internal market, the costs of exporting surplus production would be borne by the organized producers.

—The Treaty of Rome entrusted agricultural policies with safeguarding fam- ily farms, but mechanisms must be adopted that permit this to happen.

—The technological and commercial diversities of the various markets need different mechanism to deal with them, and an effort must be made to harmonize the subsidies given to different products.

—Production is no longer the farmers' only task: they are the guardians of rural territory and natural space that need oversight that demand discipline and work. These tasks must be organized and remunerated. You must accept them; even claim them, contrary to your predeces- sors who heckled me because I said that farmers must think of them- selves as the "gardeners of nature".

—Sold as they are, or processed, certain products attain a level of quality that is defined by an appellation, guaranteed by professional regulations, and also protected against unfair competition. Their specificity is linked to the small size of their territory, but they need organizations in order to get known everywhere.

—To guarantee the safety of practices and the quality of products, we obviously need to establish an organism that certifies agricultural practices and food products, and that has all the powers to investigate, and also to impose sanctions.

—To make certain that the European policy we have sketched is relevant and has legitimacy to ensure everyone's adhesion that is needed to make this happen, a convention between the representatives of all the constituents of society must define the principles that found it, along with the agricultural-global social pact that was suggested in your report.

—The time is ripe to negotiate an international agreement on these bases, because the United States has put in place a policy in order to deal with problems that are much like your own. The developing continents also must deal with problems of migrating masses of peasants that the cities are not expecting.

—There should be no European agriculture and food-supply policy that is not linked with a policy of aid to development and the fight against hunger.

—For the policy to be stable and methodically adapted to scientific, technological, economic and environmental evolutions, a permanent mechanism of evaluation and a programmed review every ten years must be part of its very definition.

—The definition of an agricultural, food-supply, rural and environmental policy is not only your concern. It cannot be worked out without your participating both in your roles as professionals and as citizens.

What more do I have to tell you? Why shouldn't I tell you a story? This happened in 1962 or 1963 in France, in Parthenay, or in Bressuire. Coming directly from Brussels by helicopter, I was greeted by an immense crowd that was assembled in a meadow, with a very high stage from which I was to give a speech. I had to explain what the Germans, the Italians, the Dutch, the Belgians, and the inhabitants of Luxemburg were doing in the domain of agriculture. As the crowd was made up mainly of cattle breeders, I concentrated most of my speech on milk. After I had just finished a sentence, a powerful voice cried out from the foot of the stage and up to the microphone: "You're not going to teach us how to milk a cow!" This remark was overheard by the three or four thousand people present, and it prompted laughter and enthusiastic cries that left me with time to reflect. Interrupting the hubbub, I soon replied: "Sir, you know how to milk your cows, and set out milk cans on the side of the road. The factory makes butter and cheeses but, afterwards, we must help you sell your excess production. Isn't this true?" This also provoked laughter and enthusiasm.

This scene could not happen today, because peasants have learned that they are dependent. Yet, I can still imagine an analogous scene. I am in the large room at a Chamber of Commerce. I am addressing a group of farm-produce industrialists and merchants. During the debate, an arm is raised. A person with horn-rimmed glasses and wearing a tie says to me: "Mister Minister, you don't know enough about our industry to have the right to speak about it as you're doing now". And I answer: "Mister President, you know how to grow and sell fresh produce, butter and cheeses, but you are undoubtedly not aware of the problems that those who deliver their milk to you have. I am here to speak about this because if you continue along the same path we will hardly need any more peasants and there will be too few farmers to infuse life into our territories. You must accept mediation to take place between you and your suppliers because this will ensure the long-term. Their own, as well as yours".

Society at large no longer understands you. It tends to blame you for "mad cow disease", pollution, the subsidies it gives you; it tends to believe it can get on without you. Society's need for nature, its tender feelings for the land, compensate for the criticism it hands out, but don't expect it to take the initiative on the idea you have launched, on the offer of a Pact that would link all of you with society, each party committing itself to satisfying the needs of the other. Take up this idea yourselves, once again; define the contents of the Pact; how it should be negotiated, along with the means that will ensure it is respected. Then, society will have to respond; because it knows all too well that its criticisms are unjust and that the expectations it expresses are legitimate. Do this in such a way that your intervention becomes a political act, whose objective is the Common Good.

Conclusion

Life, People, Land, Time … Politics

Agriculture is more than agriculture. It sustains and accompanies *Life*. It feeds People and is the guardian of the Land. Its fundamental meaning can be found only in the long-term. Feeling threatened, it appeals to the political sector to combat the indifference of society, the furies of Science and the Market. Once again, I would like to underscore the questions I have raised.

People: How can we produce enough food and distribute it so that all human beings can eat their fill, how can we restrain the demographic explosion that is threatening our balances? How can we temper the intolerable waste of wealthy civilizations? What will we do when the time comes, when those who are currently resigned to going without will no longer tolerate this state of affairs? How, armed with scientific progress, can humans support other humans without claiming they are creating a post-human? How can human society remain master of its own destiny?

The Land: Is being stripped bare by the human species who, under the pretext of bending nature to its needs, its curiosities and its discoveries, takes much more from the land than it can give back to it, thereby compromising its balances, its climates, the diversity of its soils, and the species living there. The land is threatened. How can we make it possible for it to remain, living, fertile, and diverse; and at the same time nourishing, especially in the future? How can humanity be respectful of the diversity and the unity of its cradle and the environment that nourishes it, and also of it own duration?

Time: Because of our immoderation, there is a growing contradiction between advances made in so many areas, versus the Progress the human species is making towards its ultimate fulfillment. The rhythms that are sweeping us along are an obstacle to a symbiotic relationship between human beings and their environment. We concentrate on the moment and this deflects our attention from the long-term. In our enduring lives, from birth to death to birth, a long-term view is our collective responsibility. How can we teach the need to take the long-term view to the first generation of those who have the capacity to destroy what we need to preserve?

Life and Society need Science and the Market; but cannot allow themselves to be subject to their complicit dynamics. The only way we can establish a moderator, a mediator is by reinventing *the Political Sector*. There is no other solution because we have not yet come to the end of history. The "Political" here is something that is claimed and defined by all, and it is based, at the same time, on long standing realities, and new realities defined in terms of a long-term future. How can we make it possible for the "Political" to be the architect and the guarantor of the Common Good in France, and in Europe? How can we imagine and establish a Common Good for the World?

The insane and yet reasonable ambition of this book was to ask these questions, and sometimes to dare to suggest answers.

Appendix

Glossary

Actors

During my long journey, I met peasants and farmers, young and old, from the North to the South, from the mountains to the plains. They were all so different from one another. I met country mayors, who were trying to administer communes that were too small, and too numerous. But primary school teachers, mail deliverers, country doctors, priests, rural craftsmen are getting more and more scarce. Farms are becoming secondary residences, market towns are abandoned. I worked with civil servants: side by side, they taught agronomy, supervised the forests, cared for animals and undertook rural tasks. I saw agricultural credit, co-ops, chambers of agriculture and unions develop. A bureaucracy was born from all of this activity that helped a great deal, but often took power from the hands of farmers who were elected, devoted and wise, but were too busy with their farms. I had the privilege of meeting great leaders. They were so different from one another, always in a hurry, except in my presence because they had chosen to show solidarity. In Brussels, and in European capitals, I negotiated sportingly with my five colleagues, and with Sicco Mansholt, Commissioner in charge of agriculture. He was a hearty fellow, who was competent and courageous, clever and loyal. When, later on, I came back to Brussels as commissioner in charge of cooperation and development, I met colleagues from Africa, the Caribbean, the Pacific, the Maghreb, and I saw how easy it was to work when we have an administration that functions well and along with professional, resolute and informed leaders. I worked with researchers and teachers, and I learned from all of them.

Agricultural Products and Food Products

When milk is put in a can and leaves the farm, it becomes an agricultural product; when it leaves the factory after being transformed or bottled, it is a food product. Wheat, beets, veal or lamb are agricultural products, bread, sugar, cutlets, jams are food products. Because of transformation and distribution, the prices agricultural products represent are decreasing compared with food prices; whereas agriculture and fishing supply the quasi totality of our subsistence. The percentage of food prices in the family budget is also decreasing. As this is the case, a reasonable increase in the cost of agricultural products would not disrupt family budgets.

171

Agriculture and the Rural World

For a longtime, agriculture was a state of being; and it still is in most of the world. There is a chance that modernization will simply turn agriculture into a profession.

The rural world is what exists outside cities. For a longtime, it was almost populated only by farmers. Today, it is both deserted because of the rural drift of peasants and conquered by service activities, by tourism and secondary residences. Everything has changed. Administration is all that is left with its thirty or so thousand communes that are rural only in name. Reforms are happening, but there is not the vision to establish an administrative map that would correspond to the real and would contribute to giving life to the rural world.

Agronomy

This is a pluri-disciplinary science that analyses, orients and organizes everything that concerns agriculture; it has a tendency to become more complex as the agriculturists tasks become more diversified, as the sciences and technologies become more complex, as the balance between the agricultural and the rural changes, as rural society is confronted with new problems. Today, it spans from animal and plant genetics to sociology.

It must retain its function of being the predilection of generalists.

Bureaucracy

The daughter of suspicion, of the difficulty of defining rules and concepts, of textual pointillism, an instinct for quarrelling that is increasing, bureaucracy proliferates. To vanquish it, if ever this were possible, we do not simply have to abolish budgetary positions, but we must redefine functions and simplify procedures, learn to "write" texts. The institution of the Common Agricultural Policy in Brussels, gave rise to a General Headquarters with innumerable people, and prompted an increase in the number of positions in all the national ministries of Agriculture. Because the European Union is not guided by a political vision, it has become an unbearable bureaucracy. I found thousands of texts that informed me about the Common Agricultural Policy, and very few that helped me follow the evolution of agriculture, food-supply and the rural world.

Bureaucracy has become a power that the former Administrative structures did not have. In the firms themselves, the balance between workers, directors and stock holders is rarely found, and this benefits only the directors. Learned machines exacerbate our bureaucratic illness.

Burden of Proof

Every society is founded on values: on a "code", whether written or not, that accompanies it. Because of scientific, social, political, ecological or economic evolutions the code gets more and more complicated with new rules and exceptions. It looses its capacity to be read, and also its acceptability. In our area of concern, the market is our reference. It is an efficient mechanism to drive and regulate economic activity. But because its ends are only economical, it doesn't respond to all the needs it claims to satisfy. We must correct its errors and compensate for its failings but, because it needs to be corrected and compensated for, we must assume the "burden of proof". Each time, we must define a "need" and justify the measures this need requires.

The fundamental problem with "burden of proof" arises both in practice and ideologically. Our fulfillment is founded on the balance between society and nature, on the one hand, and the economy, on the other. These two requirements are in a perpetual state of tension. Some people give priority to the economy without neglecting society, others take the opposite tack. For the moment, the differences are not cataclysmic, but they are in the long run. Today, in a debate that is dominated by market ideology, the responsibility to take the wise decision lies with those who, in the long-term give priority to nature and to society. They must therefore assume the burden of proof. We must demonstrate and argue. But we must be given the time to do so, and those with different views have to listen to us.

Civilization

By dint of hearing people talk about agriculture in market terms, I feel a violent need to respect it as the fundamental element of our civilization. To be understood I will quote Jean Rostand, the biologist: "Everything humans have added to Humans, is what we call, in the main, civilization". If we think seriously about history, we have the right to say that no sector has brought us as much as agriculture has. From the beginning of human time. Are we aware of this?

Common Good and Common Goods

There exists a specific agricultural common good that is grounded in, and is guaranteed by, the Common Good of all of society and its territories.

Crises

This is the word we use to designate incidents, or illness that societies, politics or the economy undergo. They reveal an illness, and call for a cure. For the Chinese, a crisis reveals a transition, a mutation; it is a painful passage but a promising one. We have to take advantage of this and learn how to invent in order to get beyond this. Hurrah for the crises that make those responsible question what they do. Unfortunate are those who shut their eyes and turn away. In our day the blind proliferate.

Cultivator, Farmer, Peasant, Multi-active, Pluri-active

The word cultivator refers to a profession, it is expressive and neutral. As it only retains the function of production, the word farmer expresses the demands of a remarkable adventure.

The multi-actives are those who develop their activity in multiple domains, and willingly become the artisans, the transformers of their own productions. The multi-functional, fulfill functions other than the production of agricultural products. The pluri-actives are those who are part time farmers and exercise non-agricultural activities. This sundry group often constitutes the still living tissue of rural zones in difficulty.

The word peasant, for me, is richer in meaning, because it evokes an activity, a state, duration and roots. When I use it here, I do so with respect because those who keep this name bring life to countries, and to country-sides. There are more than two billion "peasants" in the world. They are threatened by evolution, and the cities cannot take them in. A new form of modernity is evolving that shapes a different type of peasant, but also different from what the need to produce more wanted to make them.

Ecology and Environment

The environment is made up of the surroundings, in particular the natural surroundings, in which activities are developed. From our bedroom to the common room, from the apartment to the building, and to the district, from the city to the region, to the Country, to the continent, and to the world, our life unravels in environments that are interlinked like Russian dolls. From the smallest to the largest, every case of negligence, of aggression or destruction, impacts on the whole. No catastrophe has been more spectacular than the melt-down of the nuclear reactor at Chernobyl in the Soviet Union. However, the sum of the millions of petty crimes "against nature" that hundreds of millions of human beings, factories or agricultural operations, commit each day is more threatening to the future of the Planet than Chernobyl.

The ecology and ecologists are the watch dogs, the experts, the censors who struggle against these aggressions. They tell us how to repair damages, and how to avoid them. In the area of agriculture we cannot give nature up to technological aggression, and we cannot be satisfied with the establishment of protected zones.

With the help of science and disciplines, all of agriculture must respect nature and care for all of it. This is undoubtedly costly, but how much has contempt for the environment already cost us?

Food

Food is what we have on our plates, or in our glasses, in our hands when we are having a snack. The quality of our food products has increased on the whole, but not how we feed ourselves. We eat any-where, at anytime, and any way. Certain products on the market are harmful: bread that is too white, sweet beverages, the popcorn we nib-ble at in front of our TVs, or at the movies. Gastro-intestinal problems are more and more serious, and the number of obese people is on the increase. Certain foods should be forbidden like tobacco has been, because they cause more harm than it does. In the public imagination, the quality of food is now threatened by animal diseases that can be transmitted to people. This preoccupation is greatly amplified by the campaigns it has incited. We have to pay attention to this, but within the bounds of reason.

The pleasure each person wants to experience in eating plays a more and more important role in his or her life. In fact, there exist two markets that give rise to different behaviors. Let's take for example the wine and bread markets. Instead of representing ten to twelve percent of our wine consumption as it did in 1950, "appellations contrôlées" today represent almost twice as much. As their sale of ordinary prod-ucts was declining in the past, bakers reinvented breads of old, rich in fibers; people are queuing up in front of their shops.

For us, food represents daily life or the good life. But there are hun-dreds of millions of human beings "for whom their daily bread is more or less weekly". We must be concerned by this. We cannot speak about food, without speaking about hunger.

Global Economy

Economists and technicians attempt to dissuade us from keeping our books. They keep their own that are both skimpy and annual. For a longtime they have confused growth and development; they elaborat-ed the concept of human development very late in the day; they hesi-tate to take into account social costs and often refuse to calculate the cost of non-intervention. Liberal economists are accountants; they refuse to apprehend the global economy and only retain what can be immediately translated into numbers. Having adopted this attitude, they refuse to pay for the damage they cause, or they rise up against the sometimes huge costs of their failures to see events. If we try to be responsible for the unpredictable as well as the predictable, it is clear that we are wasting our time or we are taking on too much. It is also clear that we come up against a dead end, if all we do is count our money. This is the case not only for agriculture but for most other domains. An agricultural policy cannot concentrate only on production; there is all the rest that has to be taken into account.

Globalization, WorldWide

If it were a question of developing worldwide exchanges, their globalization, we would only have to solve technological problems. This would be difficult because of the different situations and levels of development of countries, because of the administrative and customs systems in place. This would be difficult but it could happen through patient negotiation. For, if such negotiations were concerned with unity, peace and development, we would not end up with intolerable uniformity.

But today we are dealing with globalization, that is to say, a well put together package where: 1) globalization of trade; 2) the adoption of a model of development that suits only a few countries or regions of the world which would ruin the planet if it happened to spread; 3) a political model that does not take into account cultures or traditions; 4) a market place capable of ruining the agricultures the world needs, and will need; 5) the domination of economic and financial matters by the WTO; 6) a unilateral political ambition, coexist.

We are faced with an egotistical ideology that would destroy nature and destroy the magnificent and indispensable diversity of the world, if it became dominant.

Growth and Development

In France, the thirty glorious years were years of growth and development: they brought us greater riches and better equilibriums. We are currently going through uncertain times: our riches are not increasing and our delicate balances seem to be broken. We have acted as though the increase in the GNP of the poor countries were part of general growth and balanced development. This was certainly not often the case. We are discovering today that growth without human, social, societal, political, educational development is lost in the sand. We have favored food aid and project aid on an ad hoc basis and acted as though a country were not a living entity. Now each country is an organism: no matter how rich it happens to be, in oil for example, it will die if it does not develop globally as an organism.

Hunger, Destitution, Poverty

Hunger is the daughter of destitution, which is very different from poverty. A poor person can think of the future. The hungry person's only future consists in discovering a crust of bread.

Ideology, Utopia

Ideology bases everything on a simple and simplifying idea, and we know the price we have to pay for it. Utopia proposes a vision, that once it has been examined, tested, amended, is worth carrying out. Ideologies proliferate, utopias are very rare.

Instant and Duration, Futurology

Present and future balances are dependent on elements as disparate as world prices, and the destiny of forests: and agriculture is a case of a perfect equilibrium between the needs of the instant, and the respect for duration.

Beginning with a rigorous analysis of the facts and tendencies, but also envisaging their possible combinations, futurology designates plausible futures. It doesn't tell us what we have to do; it warns us and suggests what we should not do. It can be mistaken; but by identifying indicators, it clarifies choices before we make decisions. We are better off making a mistake than not having attempted methodically to pierce the future.

Lands and Territory

Because they are restricted in size, lands are natural, familiar, intimate, spaces; they have long histories, micro-climates, a quality of soil, landscapes, roads that have existed forever; they often have vineyards; we incessantly quarrel about them without being able to live without them. They represent eternity before our very eyes. We have to safeguard them like natural and cultural works of art. The territory is a sovereign space. It is sovereign for animals that stake it out and defend it. It is for a tribe, for a Nation that decides to occupy it alone. Territory and Nation are twins in the very first article of our Constitutions. When I look over our Territory like an old peasant looks out over his land, I experience internal conflict where tenderness and coquetry vie with the demands of history and the world.

Markets, the Market

Across the Orient, in Beijing as well as in Casablanca, in rue Mouffetard, Sarlat, Uzès, Los Angeles, and Africa, there are markets where producers bring their odd-sized but very tasty products. This is a real joy. For the last half-century, though, and for the convenience of consumers, we have had huge supermarkets where all the products that come from around the world are cleanly arranged and highly polished there before you. They await the quiet strolling of buyers who are not clients, and even less "practices". People are all in a hurry to fill their carts before paying at the check outs'; then they disappear into their cars and hurry off again. Then, there is the Rungis food market on the outskirts of Paris.

There is the market, a mechanism which promises to ensure the balance between supply and demand, world-wide. It creates a dynamics that is unlike any other, it distributes well to those who can buy. It is not at all concerned about others. We therefore have to deal with it.

Modernity

Does modernity for agriculture consist in quantitative performance? In the past I was taken in by this, but I have since changed. We must produce more but without destroying nature and without making rural society disintegrate. Modernity for agriculture is much more difficult to realize than I had imagined.

Modernity consists in a systemic, multi-criteria and worldwide approach to the contradictions that can exist between the subsistence of all and duration; that of natural and societal environments. To be modern, we have to learn to negotiate with ecologists as well as merchants. We must assume the complexity that our concern for producing has made us forget.

Productivism and Productivity

Productivism forces larger and larger factories to produce more everyday for a bigger and bigger market; it is a conquest. By making the various factors of production more efficient, productivity tends to make factories of all sizes, along with markets of varying sizes prosper.

Scarce Commodities

The market recognizes only solvent goods. How can one think about agricultural policy without thinking about other people? Drinking water is a commodity that no one can do without; to subject it to the law of the market corresponds to refusing it to the most destitute; not to sell it at its market price is to favor waste, be it agricultural, residential or domestic. You have to know when to sell it and when to give it away. But thirst and hunger are not the only needs we must satisfy. Land is a rare commodity in certain regions of the world, either because there is not much of it available, or it has been monopolized.

The law of supply and demand cannot be the only solution to problems of subsistence and survival; one has to determine how these will be tackled and resolved.

Trees and Forests

The forest taught me so many things. It helped me get over my impatience for days and seasons to go by. It taught me duration, nature without man, even in the best kept forests.

The forest is not a crowd, but a community of trees. When a road cuts a trench through it, the surrounding trees are affected. They had got used to living together in company.

One day, I would like to write an article, a book on the forest and state why even while loving it I changed the ways of publicly and privately managing it.

Unity and Diversity

The history of the world and first the history of agriculture will undoubtedly be written around the confrontation between unity and diversity. We are living through a moment of uncertainty and the world evolving around us is hesitating between an empire or an association of equal partners; between homogenized civilizations and multiple originalities; between industrial agriculture and a respect for nature and tradition. Let's stand firmly as defenders of diversity that spares us from uniformity which is synonymous with boredom and despair and, undoubtedly, with death. We were Jacobins before the Revolution, we are tempted to remain Jacobins in the vision we have of Europe. We have to get over this. We have to learn that diversity and unity are meaningful in and by their dialogical coexistence. Let's accept, let's hope, let's act so that visions and diverse realities coexist; so that, enamored with unity that will guarantee their coexistence, they will consider each other as complimentary and necessary rather than as enemies.

Only diversity that is cultivated can lead to unity in the world. Only diversity can feed the world that needs all of its agricultures to survive.

Life

Is shaped by artisans and fabricated by industry.

A farmer raises crops and animals because, day after day, following the seasons, intervening when necessary, he accompanies the seed that germinates, the bud that blossoms, the cow that gives birth, the must that ferments, the fruit that ripens, the wine that ages and which, one day, will be beyond its prime. Life that lasts and renews itself! Nature.

The Political must be the cultivator of societies and the world.

Acknowledgments

The idea to write another book on agriculture did not arise out of the blue. Concerned by the state of confusion of the world of agriculture and the rural world, by the hunger that torments hundreds of millions of people, by the doubts that have dominated public opinion because of Mad Cow Disease, and of genetically modified organisms and agricultural pollution, friends asked the old man that I am to tell them about his experiences of yesteryear and his thoughts today. I hesitated. But while I was looking for reasons not to write, I found only reasons to write this book. The turn taken by successive policies and the projected reforms distances us from what is essential. I therefore decided to raise the question concerning these policies and proposed reforms, without however getting completely sidetracked. I decided to escape from an agricultural logic and establish a dialogue between the needs of the agricultural world and those of society at large.

I consulted enlightened people in order to enrich my own reflections. I must mention all of them as each one encouraged me, warned me, informed me, corrected and inspired me. I would like to mention their names in the same order they appear in my appointment book: Jean Pinchon, Bruno Guichard, José Rey, Raymond Février, Denis Hairy, Pierre Alphanéry, Jean Moulias, Michel Boulet, Marcel Mazoyer, Laurence Tubiana, Jean Christophe Kroll, Lucien Bourgeois, Bernard Chevassus Au Louis, Henri Nallet, Pierre Yves Guiheneuf, Bertrand Vial, Laurence Roudart, Claude Aubert. But I would like to mention especially Jean-Marie Borzeix who suggested I write "I was a 'productivist'…in the past". Each one of my children brought me his or her experience and demands. Bernard Hervieu encouraged and enlightened me. I have an unforgettable memory of an interview I had with him and Philippe Lacombe. My editor, Jean-Claude Guillebaud, welcomed and guided me. I was touched by him one day when he mentioned the old great oak-tree in his family home, uprooted by the storm of 1999 that devastated most of the regions of France. With pen in hand, attentive and demanding, he was my first reader. This work would not have been completed without the help of Claude Roger, a scientist at the National Institute for Agronomical Research in France; I benefited from his experience; he helped me discover what I did not know and forced me to be more rigorous in my analyses and more accessible to my readers. He organized a whole series of documents that I would not have been able to consult otherwise. For a year and one half we experienced together my rediscovery of agriculture. I could not have found a more worthy companion.

To each and every one I express my gratitude and consideration.

An Old Man and the Land

Postscript

An Old Man and the Land was published in Paris at Éditions du Seuil, in January 2004. Now on the eve of its first anniversary, it will appear in English and Spanish, and later on in a paperback edition, in French. Instead of analyzing it for those who will read it in one of its new editions, I would rather discuss what happened to the book during the first year after it appeared, and also the progress I made during this same period.

For a whole year, my analyses and proposals were the object of debate and criticism. They first appeared on the Website of the French newspaper *le Monde* "lemonde.fr", and then I had my own email address «www.vieilhommeetlaterre», where debate and criticism was accompanied by a hundred or so articles. I was subject to questions during more than fifty public meetings, which were attended by over fifteen thousand people. I also benefited from questions, criticisms and suggestions made by numerous experts and researchers that enabled me to make greater progress than I had in the book. Here, therefore, is encapsulated the progress made up until now.

Will the world be able to feed the world? That is "the" question, because it involves both the farmers, whose first function is to feed people, and other human beings, whose survival and life depend on the food they have, and for whom war and peace are linked to subsistence.

Technically, the production of food depends first of all on the availability of land, water, and also capital, since modern agriculture is a "heavy industry". These three elements can become very scarce. Food production; also depends on scientific progress. Until now, this has been the domain of developed countries, but it will spread to other countries. But even if scientific progress continues to be made, we must be prudent because the use of certain products or techniques can create health and safety problems for people and for the environment. Hence, as far as the technical level is concerned, the prognostic is on hold. We can therefore consider the attitude to be imprudent of those in the New World who proclaim or act as though they will find a solution to all the problems. In the near future, water, or lack thereof, could very well prove them wrong.

Two ideas were developed, when "commercial" concerns were addressed: except through price fixing, the market cannot, ensure the distribution of "goods". This was borne out by the hundreds of millions of humans who suffer from malnutrition or hunger. But now, a serious problem of a completely different nature has arisen; the announcement of the drying up of petroleum reserves, the predicted increase of consumption (in particular automobiles), the increase in prices of gasoline, lead us to believe that land, water and capital could very well be used more profitably in other areas than in food production.

From a sociological perspective, the subsistence of the nine billion humans who will populate the world in less than fifty years, and the increase in the consumption of meat products lead us to believe that we will have to triple our need for calories during this period. We can triple current needs without really factoring in the waste that inevitably occurs when societies begin from destitution, pass through poverty, before attaining a level of actual comfort.

Politically, the world has adopted a form of behavior, and countries have put in place practices that favor large integrated farms to the detriment of family agricultural operations. Deliberately or unconsciously, the world has therefore chosen to make a very large number of farmers disappear across the planet; whether they happen to be in the most populated countries in the world, or in countries that depend only on agriculture to survive.

To cut to the chase, the world could feed the world if it really wanted to, and all it would have to do is change certain rules and practices. Since it is prisoner today of an ideology and a dynamics, the world will change only if it is forced to, and then it will be too late.

+

We must analyze European agricultural policy from this perspective. A policy is nothing more than a set of coherent measures taken by an authority in order to obtain certain results. Let's look more closely at the history of European agricultural policy. In 1961, the six countries that made up the European Community wanted to become self-sufficient with respect to food production and consumption; and also wanted to balance their external exchanges. They wanted to share the "green power" that the United States had a tendency to monopolize. France played an important role here, and wanted to compensate for the industrial advantage Germany had gained from its reconstruction. No matter what the motives and hidden agendas happened to be, agricultural policy was responsible for making the Community the second largest agrofood exporter in the world.

Since it had met its goals, this policy should have been changed. All the Community did was to modify it from time to time, but only on the margins; so the result was that the policy created a greater and greater draw on the budget. It also became very unpopular among rival countries, as well as among poorer nations that accused European agriculture of unfair competition. This was followed by the advent of free trade policy, then the globalization of trade; and the concept of globalization soon began to ignore the diversity of natural regions, of civilizations that developed there and food security that were still the responsibility of national and regional political entities.

By unlinking the attribution of subsidies from actual production, the latest reforms suggest that all the Union has to do is to give the "farmers" a lifelong subsidy at the very beginning, and totally ignore their nourishing function that makes them the breadbasket of the world. In fact, it is as though it made more sense to give the responsibility for feeding the world to a few hundred thousand mega farms, and completely ignore the billions of farmers who, once they are no longer useful, will simply abandon the land and migrate to cities, factories and offices, where no one wants them. No one seems to have bothered to take into account the rural exodus that is happening in China, Africa and India, in Latin America, Indonesia but also in East-European countries, where the former State farms are rented at low prices to foreign companies that exploit a willing workforce, and are now beginning to emerge as major exporters on the world market.

Because of its categorical and uniform nature, the policy adopted by the European Union constitutes a catastrophic threat to traditional agricultures that, instead, should have been modernized. This is particularly the case in countries currently joining the European Union that refuses to maintain the agricultural and rural model which remains an important element of its civilization. We are facing a phenomenon where agricultural activity is being merchandized and deterritorialized, and where productive rural and cultivated spaces are being turned into deserts. It seems that this is happening without anyone in power worrying about what could possibly happen in the future, because the subsidies that are unlinked from production are simply temporary solutions to a crisis that is profoundly affecting agriculture.

Some people conjure up the period in history when Europe had to close its mines and foundries, and they simply state that doing the same to farms is really much ado about nothing. They do not, however, take into account either the individuals concerned, or the perishable nature of agricultural products. Forgetting the industrial lands that cities have swallowed up, they don't ask who will nibble away at the billions of hectares that are threatened by abandoning the farmland.

There are some exceptions: in Normandy and Brittany, there exist some farms that, thanks to real pastures and a few subsidies, can compete with hydroponic agriculture and pen-raised stock breeding. Have we chosen to produce in the "least possible natural way", at the very moment when fears about food production and safety are being raised?

The alternative between the choices we face does not oppose "extensive farms" and traditional peasant farms. It actually opposes two modern methods of production and of ensuring the subsistence of all human beings; one of which depends only on the market, the other also focuses on it, but does not forget nature and societies. Now, an effective agricultural policy must deal squarely with market needs, as well as the needs of society and of the land. Why do distinguished economists, who bombard us with their literature, only take into account the market economy to the detriment of the global economy that has to include in its ledgers society's and nature's costs, profits and losses.

+

The market must be regulated because it makes the co-existence of family and market agriculture, difficult, if not impossible. On this very point, we should ask ourselves if the land and societies of an exclusively market agriculture would last long worldwide? We now know that nature is no more inexhaustible than the water levels are, which we exploit without worrying about the future, and which we pollute, thereby making them unsafe. The example of Brittany, with its extensive pig farming that has polluted groundwater, is not unique in the world in our times. They are trying to repair today what they spoiled yesterday.

The dynamics of the market economy is, in the main, founded on savings and investments. We should consider the world as a collective enterprise that has to be maintained. However, in order to maintain it we must "at all costs" safeguard nature. We are not the owners but we are the usufructuaries of the world. There is no use in handing nature down on to our successors as a prosperous organism that is suffering from an incurable illness.

Market and guardian can contradict each other at a moment's notice; but they are accomplices in the long-term overseeing of our common goods. We must go to the bitter limits and treat polluters, those who exhaust our resources, with the slogan that has become famous in France "those who destroy must be those who pay".

+

This is necessary in the United States, in Europe, everywhere, but especially in Africa or in the Eastern European countries. Anyone who visited Romania thirty years ago will remember, because one really cannot forget, the grand plans of the State where pig farmers in white lab coats were raising tens of thousands of pigs all together. These farms were all as clean as real laboratories. On returning to Bucharest, one encountered hundreds of poor people with wagons pulled along by miserable nags, where the eternal peasants were bringing goods to sell at a very poor market. This is happening today, where the old collective farms are rented at very low prices to companies that originate in the rich Western world that take advantage of a not too demanding workforce and flood the world market with great quantities of foodstuff. But the wagons and nags remain. Neither the new agricultural policy, nor the State looking for goods to export, will actually take the trouble to invent modern-farm agriculture.

In Africa, a foreign visitor to a magnificent banana plantation, asked to use a helicopter to fly over the property. He was struck by the fact that the plantation in question appeared as a highly cultivated and bright green space that contrasted heavily in the middle of a dark and immense wilderness. The development of Africa, the end of misery and hunger, peace, will not come about because of a few prosperous plantations, but will only occur when a peasant population will have been given the know-how and the means to get up on its own two feet and walk.

No, this is not a question of day dreams, but of an actual awareness of the resignation of the rejects of modernity. They are in the process of overcoming their fatalism. To accompany the development of the immense territories lying fallow and the great number of the destitute in the world amounts to taking out two insurance policies in the future: peace that is so precarious, and the certainty of seeing a clientele develop, because to develop is to become a buyer in becoming a producer.

+

But, even though business is used to making forecasts and working out strategies and programs, it still hasn't understood, and refuses to accept that the communities of the world as well as governments do the same thing. This undoubtedly explains the difficulty we encounter in trying to work out and respect a true form of governance. This is an example of the negative effects of excessive competition. At the end of World War II, victorious, the United States dominated the world agro-food markets. The Economic European Community that was convalescent; then had the ambition to balance its external payments, that is to say to import less and to export more. Then the two great economic entities attempted to share this "green power". Any means, or almost any were valid.

But the world has changed, and world needs have increased. Especially, new countries have become exporters. In the throes of development, having the advantage of new and cheap land, a docile and undemanding work force, excellent natural conditions, these new countries fostered legitimate aspirations. They put an end to the dominance of the two great powers and brought about lower prices; forcing the two powers to come to the aid of their agricultures that were in danger. If we take into account the number of "newcomers", their production

capacity and their increasing demands, along with the World Trade
Organization that was being established; then it was inevitable that confrontation became more acute; especially since some of the new actors
were poor and needed to export to sustain their own development.

Consequently, the European Union and the United States with their
agricultural infrastructures appeared in the courts of international communities. Consequently, the question arose of whether they should
abolish all aid to farmers, or all aid to production itself. This gave rise
to a real comedy because, in suppressing aid to production, and maintaining the ability to attribute aid to farmers; it is clear that if they succeed in helping their agricultures they will maintain their ability to have
an impact on world markets. And they will use this; especially since
their major businesses have acquired vast landholdings that will impact
on their own internal markets. Everything leads us to believe that in
pleading for their own food safety, the "two great powers" will in no
way encourage their exports. If they did cease to be competitors, each
one could practice an "internal agricultural policy" that suited their natural, economical and social conditions, without troubling the world
markets. This would be a reasonable solution for everyone concerned.
But this seems to be politically impossible. Unless one favors extreme
solutions, it would be much better to encourage the advent of the time
of reason, finally.

In the long-term, international negotiation cannot endanger food
safety, environmental protection and proper treatment of social problems in each country that engages in it. We cannot apply the same rules
to countries that are acutely different, for historical and natural reasons.

+

This presupposes that on the whole domestic prices can guarantee the
quantities of food necessary for internal consumption, and that both
Europe and the United States will forgo all export subsidies. Therefore,
we have to ask ourselves if fiscal relief and an increase in consumer
prices can be "absorbed" by the economy and society. We have to face
up to hard and fast facts: Food costs are a smaller and smaller part of
family budgets, often less than the car. The proportion of the prices for
foodstuffs that go to farmers themselves varies between 20 and 30%,
depending on the products. Agricultural products that are subsidized
make up less than 66% of the consumer's food costs. So, we can estimate that on the whole doing away with farm subsidies should not
result in more than a 3 or 4 % increase in consumer costs. Handled properly, these costs could be easily absorbed in three or four years' time.

Along with these numbers, we should lay out the advantages of
this policy that we have roughly outlined. This would result in healthi-

er international relations, a trimming of the European budget that is, today, "burdened" by the CAP. But above and beyond this, it would result in a real awareness on the part of consumers, of what their daily bread actually costs.

+

Yet "man does not live by bread alone" and it is false to make him believe that the ways and means of production in agriculture have no impact on the environment. Competition has consequences on the ways and means of production. The latter have a decisive influence on the size of the "farms", and on the nature of rural societies. Finally, non productive and environmental or public activities can, at the same time, contribute to family budgets and have a positive impact on the environment and the "social administration" of the territory. Notwithstanding, it is difficult to see how today, in a civilization such as Europe, it is possible to think about a real exploitation of the territory, without taking into account the natural and human environment. There are so many experiments that have been done, and are currently being done, that threaten the future of a major part of the planet and rural societies.

A century ago, Jean-François Gravier published *Paris and the French Desert*, a title and a book that marked generations that were, and continue to be, very much aware of the creation of the concept and policies of "national and regional development". What he then wrote about France can be extended to the whole planet, today: If we only think about the most fertile farms with the highest yield rates, all we will end up doing is making it possible for some Chinese, Indian, Indonesian, African, Latin American, Russian but also Canadian, American, even European, thinker, researcher, writer, or film producer, to have the opportunity and feel the need to write a book or make a documentary entitled: "sixty overcrowded, raging, seething, disorganized megapolices; some immense refuges for stressed city dwellers, all located in the midst of immense abandoned spaces". Some people are obviously planning this, because all they dream about are of finding ways to do without these insufferable peasants! But more and more people are tormented by fears about food safety. Wisdom would have it that the same attention be given to the invention of new processes and new products, as well as the perfection of ways of ensuring the protection of public health that current food practices of production and distribution threaten.

+

One of the most striking phenomena of modern times can undoubtedly be seen in the instrumentalization of research by those in power, and those who have money. Research is no longer dedicated to knowledge, but to the governments or the businesses that finance it. And the human sciences themselves make it possible for governments and businesses to acquire the necessary knowledge to influence public opinion, that of electors as well as that of consumers. Newspapers publish articles praising the studies that establish…statistically that doing yoga exercises makes it possible to relieve stress victimizing business managers.

It is no longer a question of acquiring knowledge but of exercising power. This can be seen in genetic manipulations that attempt to mass "produce" inhuman beings or other monsters that could be considered useful. Since society could not trust researchers, it had to create ethics' committees. We must put in place procedures that prevent science from becoming an instrument creating dangerous or incongruous knowledge. But we must in no way prevent curiosity that characterizes each person and each animal in this world from emerging.

Because science facilitates life, nutrition, health, the environment, functioning of societies; because science is a promise, it must develop freely. But because some of its discoveries harbor real threats, society needs to define values and proscriptions; and researchers themselves must also define a deontology that ensures values and proscriptions are respected. The exuberance of research, today, along with the profusion of "progress" accompanying it, makes us hope that wisdom will be the next conquest of science. Without going further, I would like to note that there are hardly any areas of human thought that, in one way or another, are not concerned with agriculture, food, the rural world and the environment.

+

Since they are naturally linked to one another, these disciplines invite researchers and philosophers to work together. But since science is organized along disciplinary lines, it studies the living itself through the lenses of separate disciplines. This had led the organizational bodies of the medical profession and hospitals to separate generalists from specialists, the former dealing with the whole organism, and the latter with particular organs.

This is also the case in agriculture where different experts and bureaucrats approach it in isolation, and run the risk of producing incoherent solutions. This is how the unlinking of various areas of support to agriculture was invented. Because they are the purview of different administrations and disciplines that do not communicate and work with one another, they are no longer coherent. Now, life imposes that coherent actions must be taken on each soil, and each farm. Because the living is a system, that is to say an ensemble in which all the components interact indefinitely. Systemics is a discipline whereby one cannot isolate the parts but must treat each part by taking into account that it lives and functions in, by, and for the unity and equilibrium of the whole. Every infraction of this rule results in a diseased organism.

To speak of agriculture in terms of systemics, is to refuse that production be the only function that counts; it is to root agriculture in its soil, whereas the market tends to uproot it from the soil. To do this is to return, but without nostalgia, to this "eternal order of the fields", all the while considering order as the organization that makes it possible to understand the constituents and the lasting efficacy of the work accomplished by farmers.

+

I have just presented in my own terms, what I learned when my book, my analyses and my theses were subjected to the open debate that I referred to at the beginning of this section. These debates undoubtedly taught me more than my research and my writings.

The discovery of a world results from the encounter between reality and the idea we have of it. To discover, we must not travel with an empty mind. We need a hypothesis of sorts and the desire, the capacity to test it on real data. This occurred in the provinces where I was both a Prefect and a Senator, in Paris within Administration and Government, in Brussels in the midst of those who invented European agricultural policy, in Africa with heads of State and their ministers, at the National Institute for Agronomical Research with researchers, in Romania in State institutions, in Egypt with the authority having the task of exploiting the water reserves of the Asswan Dam, in California in the admirable irrigated central plain, so profuse yet also so fragile; but also in masses of books and films, in my passionate and loyal meetings with agricultural leaders. The various sites that were opened, and public meetings, helped me to understand better what I thought I understood for a long time.

I am the real winner in this adventure.

I was struck by a number of things, though: none of my analyses were radically disproved; considered from a non institutional perspective, the problem must be considered in the same way I did; — people were not necessarily won over by my propositions, but they weren't contested — farmers, researchers and politicians seem to have been overcome by a pathetic form of fatality. In spite of the arguments and the somber predictions, very few people seem to feel that it is possible to escape market logic and globalization. In France and in Europe already, the agricultural world is like a species about to disappear.

However, I won't give up. But I will "flip" my logic around. The "old man" has presented a defense and illustration of an activity, a state, a nourishing and manifold function. He did so, hoping that society at large, institutions, the agro-industrial and agro-food-producing world, would understand what farmers need to have to continue.

Yet, this is not the real problem. In Europe, in the United States, in China, in Africa, in Latin America, the workers of the land feel and believe they are condemned by the logic of the market that benefits only large landholder, industrialists and consumers. We therefore have to raise these issues with them. If things continue to evolve in the same way, the world neither guarantees our food safety, nor our environment, nor the harmonious and efficient distribution of human societies. Farmers feel they cannot make themselves heard; it's up to the political to intervene because, in essence, it is the mediator between an economic dynamics that has its virtues and a society and a land that are in disarray, today. Because the political is the mediator, it doesn't have the right to hand over the destiny of the living to the sole logic of profit. Because the Soviets founded their future on their ideology of a government run economy, it crumbled. Because the tenants of globalization that everyone seems to side with have founded the future on market ideology, by subjecting the diversity of the real to the unity of the market and the duration of the living to the performances of the instant, they are preparing us for a more serious collapse, because both elements of these two alternatives will have been misleading.

I would like to raise this issue by evoking a recent event. I had a public debate in Amsterdam on November 1, 2004, with the Dutch Minister of Agriculture, who was President of the Council of European Agricultural Ministers. This turned out to be a cordial debate, but without any concessions whatsoever on either side. The next day, a newspaper article on this meeting appeared with the title: "An Old Dreamer and a Minister with No Illusions". Thus to express the very strong liberal position of the latter, they had to suggest that he was wise to have no illusions, and to disqualify the ideas of the former, he had to be labeled an old dreamer.

Doesn't illusion reside in the believing in the eternity of nature and in the unshakeable subjection of humans to the discipline of an unjust world? Doesn't wisdom reside in contesting a dynamics that, far from all that is living, wants to build the future only on market performance?

"The Land teaches us more than all books: because it resists us"
Antoine de Saint Exupéry (*Man and His World*)